Teacher Edition

Eureka Math®
Grade K
Module 2

Special thanks go to the Gordon A. Cain Center and to the Department of Mathematics at Louisiana State University for their support in the development of *Eureka Math*.

For a free *Eureka Math* Teacher Resource Pack, Parent Tip Sheets, and more please visit https://eurekamath.greatminds.org/teacher-resource-pack

Published by the Great Minds

Printed in the U.S.A.
This book may be purchased from the publisher at eureka-math.org
BAB 10 9 8 7 6 5 4 3 2

ISBN 978-1-63255-342-3

***Eureka Math: A Story of Units*® Contributors**

Katrina Abdussalaam, Curriculum Writer
Tiah Alphonso, Program Manager—Curriculum Production
Kelly Alsup, Lead Writer / Editor, Grade 4
Catriona Anderson, Program Manager—Implementation Support
Debbie Andorka-Aceves, Curriculum Writer
Eric Angel, Curriculum Writer
Leslie Arceneaux, Lead Writer / Editor, Grade 5
Kate McGill Austin, Lead Writer / Editor, Grades PreK–K
Adam Baker, Lead Writer / Editor, Grade 5
Scott Baldridge, Lead Mathematician and Lead Curriculum Writer
Beth Barnes, Curriculum Writer
Bonnie Bergstresser, Math Auditor
Bill Davidson, Fluency Specialist
Jill Diniz, Program Director
Nancy Diorio, Curriculum Writer
Nancy Doorey, Assessment Advisor
Lacy Endo-Peery, Lead Writer / Editor, Grades PreK–K
Ana Estela, Curriculum Writer
Lessa Faltermann, Math Auditor
Janice Fan, Curriculum Writer
Ellen Fort, Math Auditor
Peggy Golden, Curriculum Writer
Maria Gomes, Pre-Kindergarten Practitioner
Pam Goodner, Curriculum Writer
Greg Gorman, Curriculum Writer
Melanie Gutierrez, Curriculum Writer
Bob Hollister, Math Auditor
Kelley Isinger, Curriculum Writer
Nuhad Jamal, Curriculum Writer
Mary Jones, Lead Writer / Editor, Grade 4
Halle Kananak, Curriculum Writer
Susan Lee, Lead Writer / Editor, Grade 3
Jennifer Loftin, Program Manager—Professional Development
Soo Jin Lu, Curriculum Writer
Nell McAnelly, Project Director

Ben McCarty, Lead Mathematician / Editor, PreK–5
Stacie McClintock, Document Production Manager
Cristina Metcalf, Lead Writer / Editor, Grade 3
Susan Midlarsky, Curriculum Writer
Pat Mohr, Curriculum Writer
Sarah Oyler, Document Coordinator
Victoria Peacock, Curriculum Writer
Jenny Petrosino, Curriculum Writer
Terrie Poehl, Math Auditor
Robin Ramos, Lead Curriculum Writer / Editor, PreK–5
Kristen Riedel, Math Audit Team Lead
Cecilia Rudzitis, Curriculum Writer
Tricia Salerno, Curriculum Writer
Chris Sarlo, Curriculum Writer
Ann Rose Sentoro, Curriculum Writer
Colleen Sheeron, Lead Writer / Editor, Grade 2
Gail Smith, Curriculum Writer
Shelley Snow, Curriculum Writer
Robyn Sorenson, Math Auditor
Kelly Spinks, Curriculum Writer
Marianne Strayton, Lead Writer / Editor, Grade 1
Theresa Streeter, Math Auditor
Lily Talcott, Curriculum Writer
Kevin Tougher, Curriculum Writer
Saffron VanGalder, Lead Writer / Editor, Grade 3
Lisa Watts-Lawton, Lead Writer / Editor, Grade 2
Erin Wheeler, Curriculum Writer
MaryJo Wieland, Curriculum Writer
Allison Witcraft, Math Auditor
Jessa Woods, Curriculum Writer
Hae Jung Yang, Lead Writer / Editor, Grade 1

Table of Contents

GRADE K • MODULE 2

Two-Dimensional and Three-Dimensional Shapes

Module Overview 2

Topic A: Two-Dimensional Flat Shapes 9

Topic B: Three-Dimensional Solid Shapes 66

Topic C: Two-Dimensional and Three-Dimensional Shapes 90

End-of-Module Assessment and Rubric 102

Answer Key 109

Grade K • Module 2

Two-Dimensional and Three-Dimensional Shapes

OVERVIEW

In Module 1, students began the year observing their world. What is exactly the same? What is the same but...? They matched and sorted according to criteria sequenced from simple to complex. Their perceptions evolved into observations about numbers to 10. "4 is missing 1 to make 5." "4 plus 1 more is 5." "There are the same number of dogs and flowers, 6."

In this module, students seek out flat and solid shapes in their world (**K.G.1**). Empowered by this lens, they begin to make connections between the wheel of a bicycle, the moon, and the top of an ice cream cone. Just as the number 4 allowed them to quantify 4 mountains and 4 mice as equal numbers, learning to identify flats and solids allows them to see the relationship of the simple to the complex, a mountain's top to a plastic triangle and cone sitting on their desk.

To open Topic A, students find and describe flat shapes in their environment using informal language, without naming them at first (**K.G.4**). In Lesson 2, they classify the shapes, juxtaposing them with various examples and non-examples. This process further refines their ability to talk about the shapes, for example, as closed or having straight sides. The naming of the flat shape as a triangle is part of that process, not the focus of it (**K.G.2**, **K.G.1**).

The same process is then repeated with rectangles in Lesson 3 and hexagons and circles in Lesson 4. In Lesson 5, students manipulate all the flat shapes using position words as the teacher gives directives such as, "Move the closed shape with three straight sides behind the shape with six straight sides." These positioning words are subsequently woven into the instructional program, at times in math fluency activities, but also throughout the entire school day.

The lessons of Topic B replicate those of Topic A but with solid shapes. In addition, students recognize the presence of the flats within the solids. The module closes in Topic C with discrimination between flats and solids. A culminating task involves students in creating displays of a given flat shape with counter-examples and showing related solid shapes (**K.G.3**).

The fluency components in the lessons of Module 1 included activities wherein students used a variety of triangles and rectangles to practice the decompositions of 3 and 4. Flats and solids will continue to be included in fluency activities in this module and throughout the year so that students have repeated experiences with shapes, their attributes, and their names. Daily number fluency practice in this new module is critical. There are two main goals of consistent fluency practice: (1) to solidify the numbers of Module 1 and (2) to anticipate the numbers of Modules 3, 4, and 5. Therefore, students continue to work extensively with numbers to 10 and fluency with addition and subtraction to 5.

The Kindergarten year closes in Module 6 with another geometry unit. By that time, having become much more familiar with flats and solids, the students compose new flat shapes ("Can you make a rectangle from these two triangles?") and build solid shapes from components ("Let's use these straws to be the edges and these balls of clay to be the corners of a cube!"). This module will allow them to bring together all that they have learned throughout the year as they manipulate shapes and their components (**K.G.4**, **K.G.5**).

Notes on Pacing for Differentiation

If pacing is a challenge, consider omitting Lessons 5 and 8. Instead, embed experiences with position words in other content areas and throughout the students' day. It is not essential that students be introduced to position words through the context of shapes.

Distribution of Instructional Minutes

This diagram represents a suggested distribution of instructional minutes based on the emphasis of particular lesson components in different lessons throughout the module.

- Fluency Practice
- Concept Development
- Application Problems
- Student Debrief

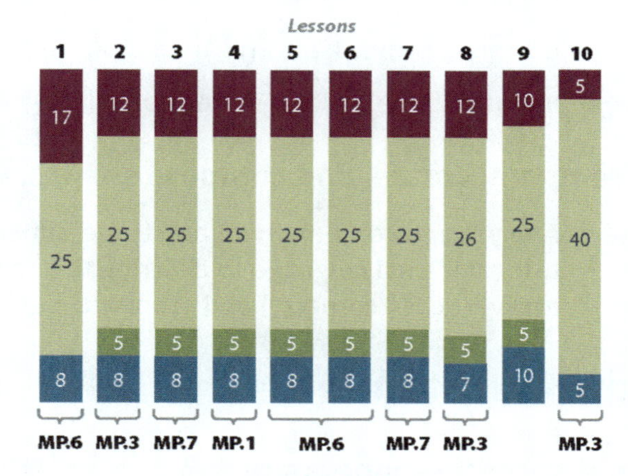

Focus Grade Level Standards

Classify objects and count the number of objects in each category.

K.MD.3 Classify objects into given categories; count the numbers of objects in each category and sort the categories by count. (Limit category counts to be less than or equal to 10.)

Identify and describe shapes (squares, circles, triangles, rectangles, hexagons, cubes, cones, cylinders, and spheres).

K.G.1 Describe objects in the environment using names of shapes, and describe the relative positions of these objects using terms such as *above, below, beside, in front of, behind*, and *next to*.

K.G.2 Correctly name shapes regardless of their orientations or overall size.

K.G.3 Identify shapes as two-dimensional (lying in a plane, “flat”) or three-dimensional (“solid”).

Analyze, compare, create, and compose shapes.[1]

K.G.4 Analyze and compare two- and three-dimensional shapes, in different sizes and orientations, using informal language to describe their similarities, differences, parts (e.g., number of sides and vertices/“corners”) and other attributes (e.g., having sides of equal length).

Foundational Standards

Identify and describe shapes (squares, circles, triangles, rectangles).

PK.G.1 Describe objects in the environment using names of shapes, and describe the relative positions of these objects using terms such as *top, bottom, up, down, in front of, behind, over, under*, and *next to*.

PK.G.2 Correctly name shapes regardless of size.

Analyze, compare, and sort objects.

PK.G.3 Analyze, compare, and sort two- and three-dimensional shapes and objects, in different sizes, using informal language to describe their similarities, differences, and other attributes (e.g., color, size, and shape).

PK.G.4 Create and build shapes from components (e.g., sticks and clay balls).

[1]The balance of this cluster is addressed in Module 6.

Focus Standards for Mathematical Practice

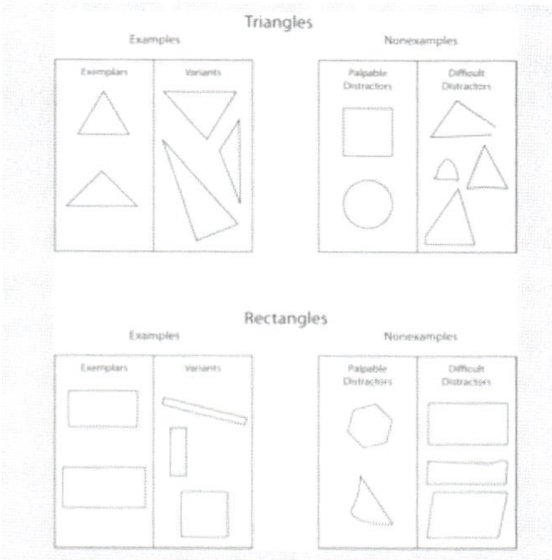

Exemplars are the typical visual prototypes of the shape category.

Variants are other examples of the shape category.

Palpable distractors are nonexamples with little or no overall resemblance to the exemplars.

Difficult distractors are visually similar to examples but lack at least one defining attribute.

MP.1 **Make sense of problems and persevere in solving them**. Students distinguish shapes from among variants, palpable distractors, and difficult distractors.[2] (See examples to the right.)

MP.3 **Construct viable arguments and critique the reasoning of others**. Students are increasingly able to use shape attributes to defend identification of a plane or solid shape.

MP.6 **Attend to precision**. Students use position words to clearly indicate the location of shapes. Also, when Kindergarten students are analyzing and defining attributes such as "3 straight sides," they are attending to precision.

MP.7 **Look for and make use of structure**. Students use examples, non-examples, and shared attributes of geometric figures in order to develop a richer concept image (Geometry Progressions, p. 6) of each geometric shape. This concept image allows for more acute discernment of the shape within the environment.

[2]This image plus further clarification is found in Geometry Progressions, p. 6.

Overview of Module Topics and Lesson Objectives

Standards		Topics and Objectives	Days
K.G.1 **K.G.2** **K.G.4** K.MD.3	A	**Two-Dimensional Flat Shapes** Lesson 1: Find and describe flat triangles, squares, rectangles, hexagons, and circles using informal language without naming. Lesson 2: Explain decisions about classifications of triangles into categories using variants and non-examples. Identify shapes as triangles. Lesson 3: Explain decisions about classifications of rectangles into categories using variants and non-examples. Identify shapes as rectangles. Lesson 4: Explain decisions about classifications of hexagons and circles, and identify them by name. Make observations using variants and non-examples. Lesson 5: Describe and communicate positions of all flat shapes using the words *above, below, beside, in front of, next to*, and *behind*.	5
K.G.1 **K.G.2** **K.G.4** K.MD.3	B	**Three-Dimensional Solid Shapes** Lesson 6: Find and describe solid shapes using informal language without naming. Lesson 7: Explain decisions about classification of solid shapes into categories. Name the solid shapes. Lesson 8: Describe and communicate positions of all solid shapes using the words *above, below, beside, in front of, next to*, and *behind*.	3
K.MD.3 **K.G.3** **K.G.4** K.G.1 K.G.2	C	**Two-Dimensional and Three-Dimensional Shapes** Lesson 9: Identify and sort shapes as two-dimensional or three-dimensional, and recognize two-dimensional and three-dimensional shapes in different orientations and sizes. Lesson 10: Culminating task—collaborative groups create displays of different flat shapes with examples, non-examples, and a corresponding solid shape.	2
		End-of-Module Assessment: Topics A–C (Interview style assessment: 2 days)	2
Total Number of Instructional Days			**12**

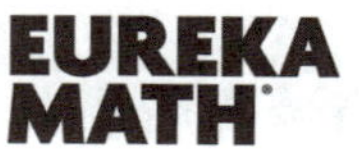

Terminology

New or Recently Introduced Terms

- Above, below, beside, in front of, next to, behind (position words)
- Circle
- Cone (solid shape)
- Cube (solid shape)
- Cylinder (solid shape)
- Face (flat side of a solid)[3]
- Flat (two-dimensional shape)
- Hexagon (flat figure enclosed by six straight sides)
- Rectangle (flat figure enclosed by four straight sides)
- Solid (three-dimensional shape)
- Sphere (solid shape)
- Square (flat figure enclosed by four straight, equal sides)
- Triangle (flat figure enclosed by three straight sides)

Familiar Terms and Symbols[4]

- Match (group items that are the same or have the same given attribute)
- Sort

Suggested Tools and Representations

- Three-dimensional shapes: cone, sphere, cylinder, and cube
- Two-dimensional shapes: circle, hexagon, rectangle, square, and triangle

---

[3]In the context of polyhedra, faces must be polygonal. However, in more general contexts, a face may be circular (such as the base of a right circular cylinder), or even irregular. It is this more inclusive interpretation of face that is used in this Kindergarten module.

[4]These are terms and symbols students have seen previously.

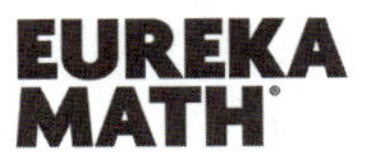

Homework

Homework at the K–1 level is not a convention in all schools. In this curriculum, homework is an opportunity for additional practice of the content from the day's lesson. The teacher is encouraged, with the support of parents, administrators, and colleagues, to discern the appropriate use of homework for his or her students. Fluency exercises can also be considered as an alternative homework assignment.

Scaffolds[5]

The scaffolds integrated into *A Story of Units*® give alternatives for how students access information as well as express and demonstrate their learning. Strategically placed margin notes are provided within each lesson, elaborating on the use of specific scaffolds at applicable times. They address many needs presented by English language learners, students with disabilities, students performing above grade level, and students performing below grade level. Many of the suggestions are organized by Universal Design for Learning (UDL) principles and are applicable to more than one population. To read more about the approach to differentiated instruction in *A Story of Units*, please refer to "How to Implement *A Story of Units*."

Assessment Summary

Type	Administered	Format	Standards Addressed
End-of-Module Assessment Task	After Topic C	Interview with Rubric	K.MD.3 K.G.1 K.G.2 K.G.3 K.G.4
Culminating Task	Lesson 10	Collaborative Project	K.MD.3 K.G.1 K.G.2 K.G.3 K.G.4

[5]Students with disabilities may require Braille, large print, audio, or special digital files. Please visit the website www.p12.nysed.gov/specialed/aim for specific information on how to obtain student materials that satisfy the National Instructional Materials Accessibility Standard (NIMAS) format.

Topic A

Two-Dimensional Flat Shapes

K.G.1, K.G.2, K.G.4, K.MD.3

Focus Standards:		K.G.1	Describe objects in the environment using names of shapes, and describe the relative positions of these objects using terms such as *above, below, beside, in front of, behind,* and *next to*.
		K.G.2	Correctly name shapes regardless of their orientations or overall size.
		K.G.4	Analyze and compare two- and three-dimensional shapes, in different sizes and orientations, using informal language to describe their similarities, differences, parts (e.g., number of sides and vertices/"corners") and other attributes (e.g., having sides of equal length).
Instructional Days:		5	
Coherence	**-Links from:**	GPK–M2	Shapes
	-Links to:	G1–M5	Identifying, Composing, and Partitioning Shapes

Students began the year, in Module 1, developing number concepts by observing their world. Now, they begin to develop spatial reasoning and geometric concepts by experiencing flat and solid shapes in their world. This module examines how shapes and objects are similar to or different from one another with respect to orientation and relative positions to objects.

In Lesson 1, students use the informal language of their everyday world to name and describe flat shapes without yet expressing mathematical concepts or using the vocabulary of geometry. At this point, students are not yet able to consistently distinguish between examples and non-examples of different groups of shapes such as triangles, circles, squares, rectangles, or hexagons. At this stage, a figure is a square because it looks like a book; another figure is a circle because it is round like the wheel of a car. Students make these observations without explicitly thinking about the attributes or properties of squares or circles.

In Lesson 2, students build on their experiential learning by relating it to the mathematical concepts and vocabulary of geometry, allowing them to enhance their experiences of shapes. Students begin to classify three-sided shapes by identifying them as examples of a triangle. Using various examples and non-examples of triangles, they sort and classify different shapes as examples of a triangle or not a triangle. Having learned to identify shapes as triangles, they explain their decisions about classifying some shapes as triangles and other shapes as not triangles by focusing on common attributes or properties of the shapes they have identified as triangles.

Lessons 3 and 4 continue the work of Lesson 2 in the same vein by identifying shapes as rectangles, hexagons, or circles. In Lesson 5, students communicate about the relative position of shapes by using terms such as above, below, next to, beside, in front of, and behind.

A Teaching Sequence Toward Mastery of Two-Dimensional Flat Shapes	
Objective 1:	**Find and describe flat triangles, squares, rectangles, hexagons, and circles using informal language without naming. (Lesson 1)**
Objective 2:	**Explain decisions about classifications of triangles into categories using variants and non-examples. Identify shapes as triangles. (Lesson 2)**
Objective 3:	**Explain decisions about classifications of rectangles into categories using variants and non-examples. Identify shapes as rectangles. (Lesson 3)**
Objective 4:	**Explain decisions about classifications of hexagons and circles, and identify them by name. Make observations using variants and non-examples. (Lesson 4)**
Objective 5:	**Describe and communicate positions of all flat shapes using the words *above, below, beside, in front of, next to*, and *behind*. (Lesson 5)**

Lesson 1

Objective: Find and describe flat triangles, squares, rectangles, hexagons, and circles using informal language without naming.

Suggested Lesson Structure

Fluency Practice	(17 minutes)
Concept Development	(25 minutes)
Student Debrief	(8 minutes)
Total Time	**(50 minutes)**

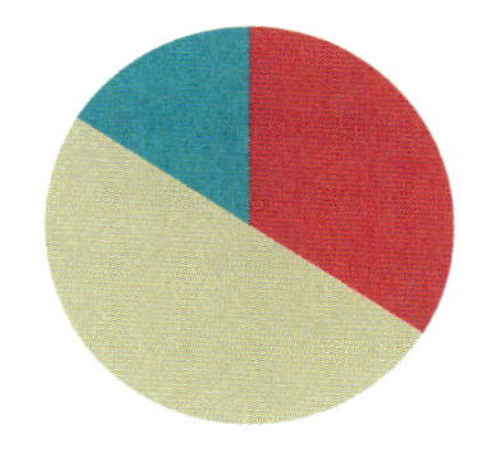

Fluency Practice (17 minutes)

- Making 5 with 5-Group Mats **K.OA.1** (6 minutes)
- Draw More to Make 5 **K.OA.3** (8 minutes)
- 5-Group Hands **K.CC.2** (3 minutes)

Making 5 with 5-Group Mats (6 minutes)

Materials: (S) 5-group mats (Fluency Template 1), 5 linking cubes

Note: While students are working with geometry, the fluency goal throughout Module 2 will be to maintain and further develop number concepts to 10 (see Fluency Practice note in Kindergarten Module 1 Lesson 1).

T: Touch and count your cubes.

S: 1, 2, 3, 4, 5.

T: Touch and count the dots on your mat.

S: 1, 2, 3, 4, 5.

T: Our job is to make 5. Put 4 cubes on the dots of your mat. (Check to see that students place the cubes from left to right without skipping any dots.) Raise your hand when you know how many more cubes to make 5. (Wait until all hands are raised, and then signal.) Ready?

S: 1.

T: We can tell how to make 5 like this: 4 and 1 make 5. Echo me, please.

S: 4 and 1 make 5.

Continue working through the decompositions of 5 in a systematic way. As students begin to demonstrate mastery, scale back the amount of guidance: "Show me *X* cubes; say the number sentence."

Draw More to Make 5 (8 minutes)

Materials: (S) Draw more (Fluency Template 2)

Note: Go over the answers, and direct students to energetically shout "Yes!" for each correct answer.

After giving clear instructions and completing the first few problems together, allow students time to work independently. Encourage them to do as many problems as they can within a given time frame.

5-Group Hands (3 minutes)

Materials: (T) Large 5-group cards (5–7) (Fluency Template 3)

T: (Show the 6-dot card.) Raise your hand when you know how many dots are on top. (Wait until all hands are raised, and then signal.) Ready?

S: 5.

T: Bottom?

S: 1.

T: We can show this 5-group on our hands. Five on top, 1 on the bottom, like this. (Demonstrate on hands, one above the other, as shown to the right.)

S: (Show 5 and 1 on hands, one above the other.)

T: Push your hands out as you count on from 5, like this. 5 (extend the top hand forward), 6 (extend the bottom hand forward). Try it with me.

S: 5 (extend the top hand forward), 6 (extend the bottom hand forward).

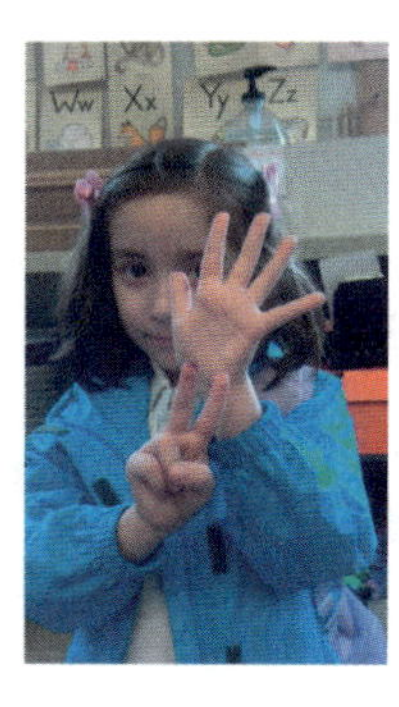

A student demonstrates 7 as 5 on top and 2 on the bottom.

Continue with 5, 6, and 7, steadily decreasing guidance from the teacher, until students can show the 5-groups on their hands with ease.

Concept Development (25 minutes)

Materials: (T) Large cutouts of each shape (to be affixed to the board with tape) (Template)
(S) Clear bag containing smaller cutouts of various shapes (all of one hue to limit distractions from variation in color), blank side of Problem Set affixed to clipboard, pencil, real or toy magnifying glass (if available)

Suggestions for shape cutouts are pictured as follows but need not be limited to these. Be sure to include, at minimum, a triangle, circle, square, rectangle, and hexagon for discussion purposes.

Note: Today's lesson focuses on the attributes of the shapes but *not* their specific names. Assure students that tomorrow's work will include naming the shapes since many may be very eager to share their knowledge.

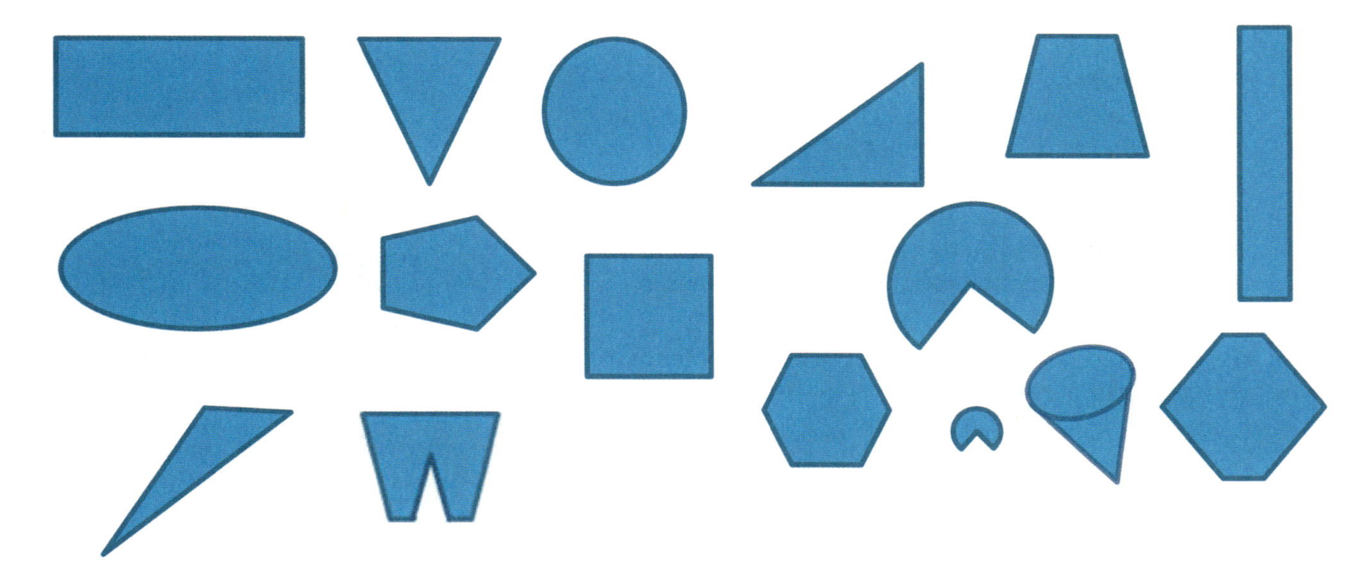

T: You have a mystery bag! Open your bag, and carefully shake out the surprises inside. What do you see? (Give students a moment to explore the contents of the bag and discuss with their friends.)

S: Different shapes!

MP.6

T: (Select a shape from the bag.) Look at my shape. Can you find the one that looks like mine? (Affix the shape to the board.) Tell me about the shape. (In order to encourage a discussion purely about geometric attributes, select one of the more unusual cutouts to begin.)

S: It is round. → But, it is pointy! → It has a piece missing. → It has three sides.

T: I like your observations! (Write student responses on chart paper, and continue the exercise with the rest of the shapes, encouraging students to verbalize attributes such as corners, curves, straight lines, number and length of sides, "missing pieces," etc.)

T: Arrange your shapes on your desktop. Do they have anything in common? (Responses will vary.) Now, bend down so that you are looking across the edge of your desktop. Can you see your shapes now? Are any of them sticking up?

S: We can't see them. → They are all flat!

T: Yes, they do have that in common! These are all **flat shapes**. Put your shapes back in the bag.

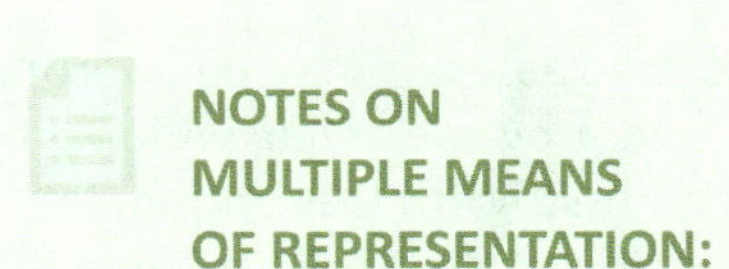

NOTES ON MULTIPLE MEANS OF REPRESENTATION:

English language learners benefit from having the words *curved, straight, pointy, round, sides*, and other attributes introduced before the lesson so that they can participate in the discussion with the class. After introducing them, post the vocabulary on the word wall with visuals so that students can refer to them.

T: It's time to play shape detectives! Detectives need to have special equipment, so I am going to give you and a partner a magnifying glass to use if you need it. You are going to go on a shape hunt around the room. Whenever you see an interesting shape, tell your partner about it, and draw it on your paper. Take your bag of shapes with you to use as clues. Maybe you will see some shapes in the room that match shapes in your bag!

S: (With partners, search for shapes, and re-create them on their clipboards.)

T: (After five minutes, call students back to their seats.) Does anyone want to share one of the shapes they found? Tell us about it! (Allow time for sharing and discussion.)

T: Maybe you will find more shapes to add tonight. Turn your Problem Sets over so that we can do some shape coloring and matching.

NOTES ON MULTIPLE MEANS OF ENGAGEMENT:

Push students working above grade level by asking them questions and assigning activities that engage thinking at higher levels. "What would that shape look like if it was not flat?" "Can you make a picture of that shape but make it so that it is sticking up?"

Problem Set (10 minutes)

Students should do their personal best to complete the Problem Set within the allotted time.

For some classes, it may be appropriate to modify the assignment by specifying which problems students should work on first. With this option, let the purposeful sequencing of the Problem Set guide your selections so that problems continue to be scaffolded. Balance word problems with other problem types to ensure a range of practice. Assign incomplete problems for homework or at another time during the day.

In this Problem Set, all students should begin with sorting the shapes that clearly have or do not have curves and possibly leave any questionable shapes to the end if time permits.

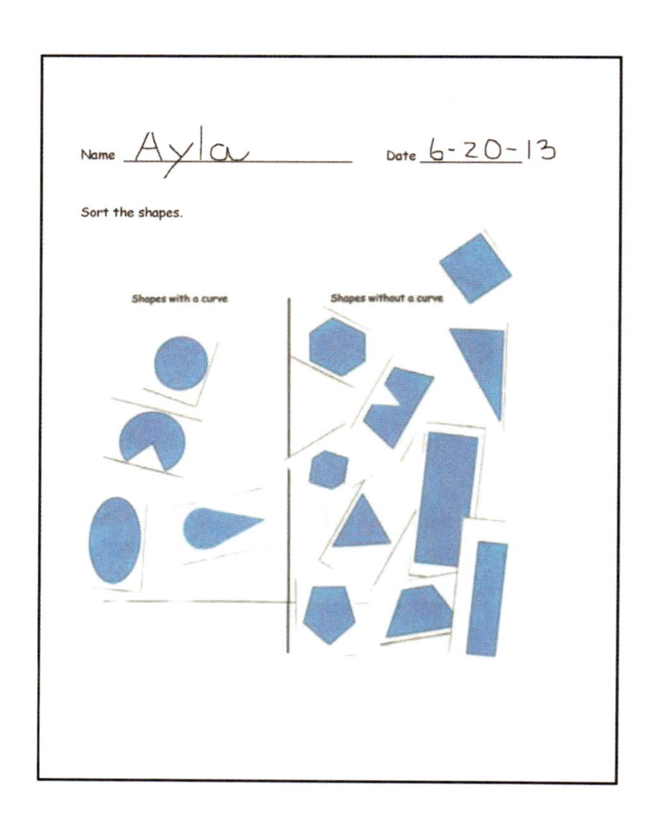

Suggestions for other ways you may ask students to sort are listed below:

- Shapes that have curves and sharp points.
- Shapes that have only curves.
- Shapes that have four or fewer corners.
- Shapes that have four or more sides.

Student Debrief (8 minutes)

Lesson Objective: Find and describe flat triangles, squares, rectangles, hexagons, and circles using informal language without naming.

The Student Debrief is intended to invite reflection and active processing of the total lesson experience.

Invite students to review their solutions for the Problem Set. They should check work by comparing answers with a partner before going over answers as a class. Look for misconceptions or misunderstandings that can be addressed in the Debrief. Guide students in a conversation to debrief the Problem Set and process the lesson.

Any combination of the questions below may be used to lead the discussion.

- Which objects did you sort that were curved? Which objects did you sort that were not curved?
- Which **flat shapes** were the hardest to sort? Why?
- Explain to your partner which shapes you drew on the back of your paper. Can you think of other objects around you that have these same shapes?
- What new (or significant) math vocabulary did we use today to communicate precisely?
- How can you tell about each shape without using the shape's name?

Homework

Homework at the K–1 level is not a convention in all schools. In this curriculum, homework is an opportunity for additional practice of the content from the day's lesson. The teacher is encouraged, with the support of parents, administrators, and colleagues, to discern the appropriate use of homework for his or her students. Fluency exercises can also be considered as an alternative homework assignment.

Name ______________________________ Date ______________

Sort the shapes.

Shapes with a Curve	Shapes without a Curve

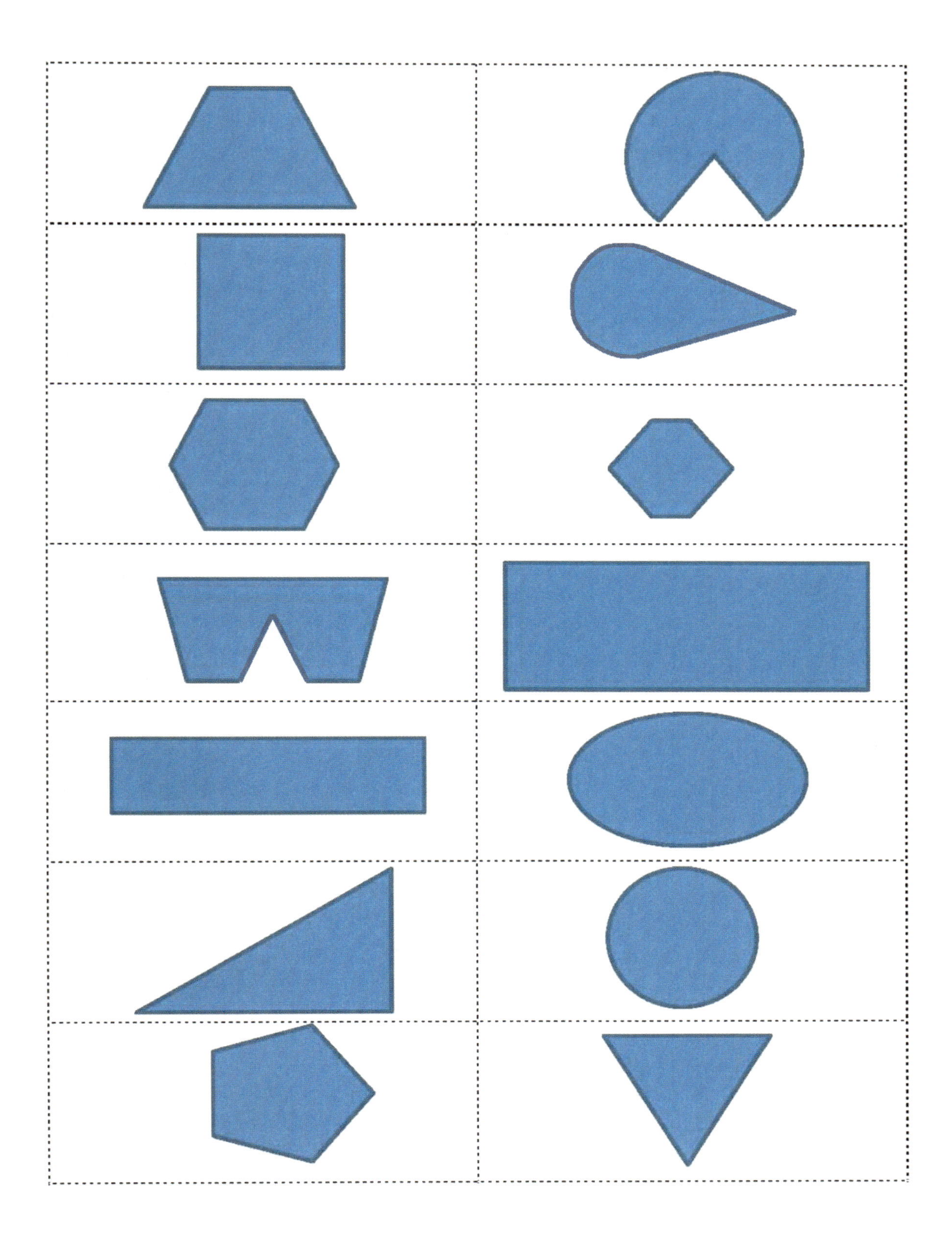

Name ______________________________ Date ______________

Draw a line from the shape to its matching object.

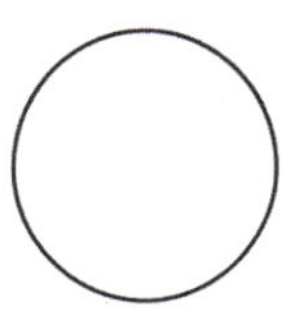

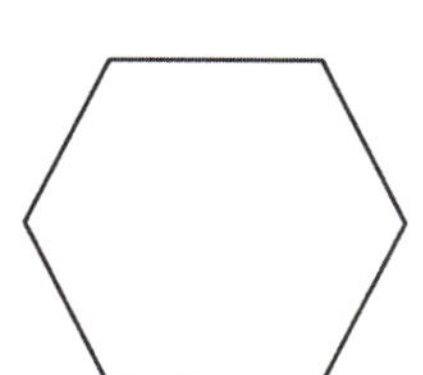

●	●	●	●	●

●	●	●	●	●

●	●	●	●	●

●	●	●	●	●

●	●	●	●	●

5-group mat

Draw more to make 5.

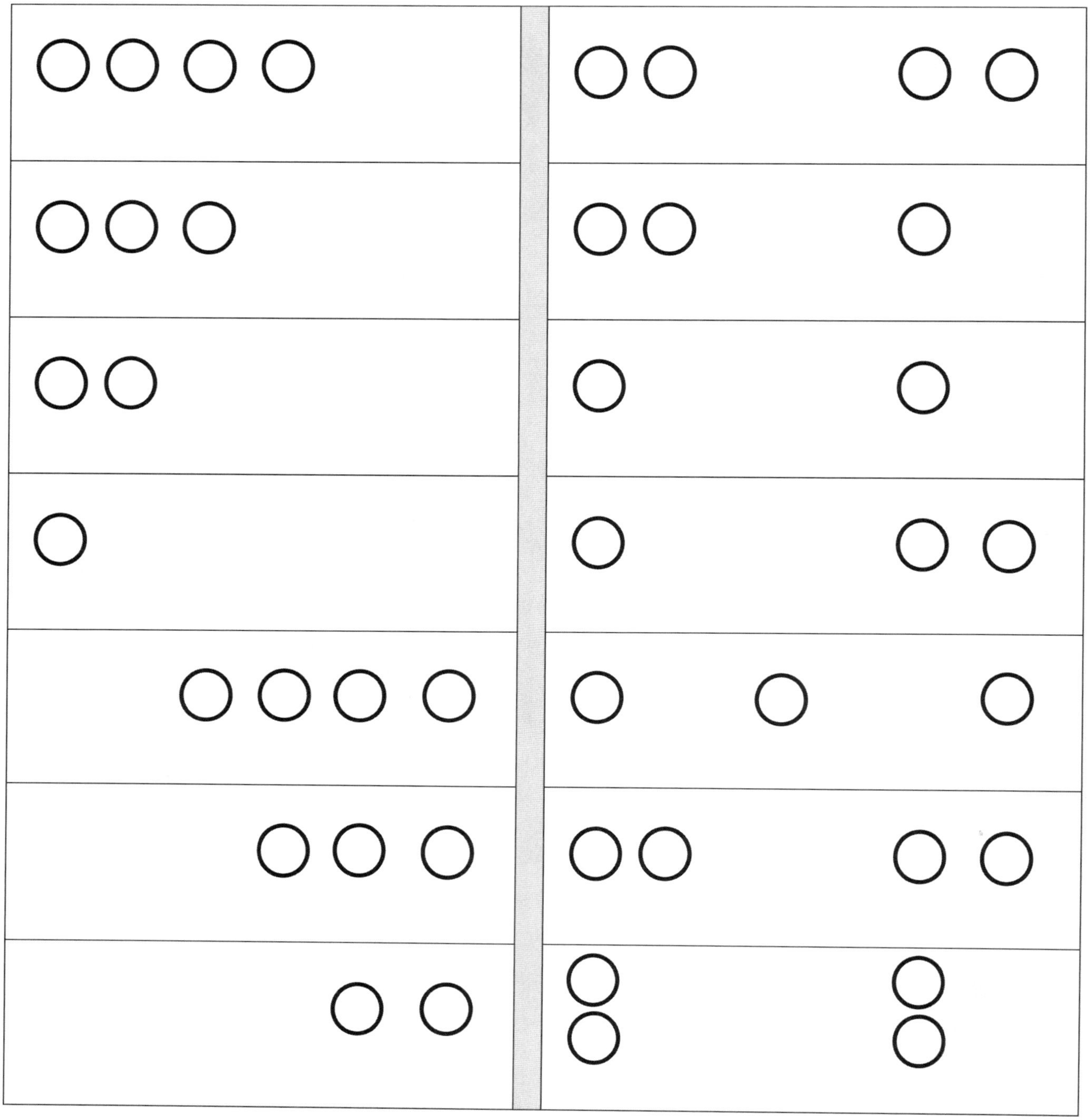

draw more

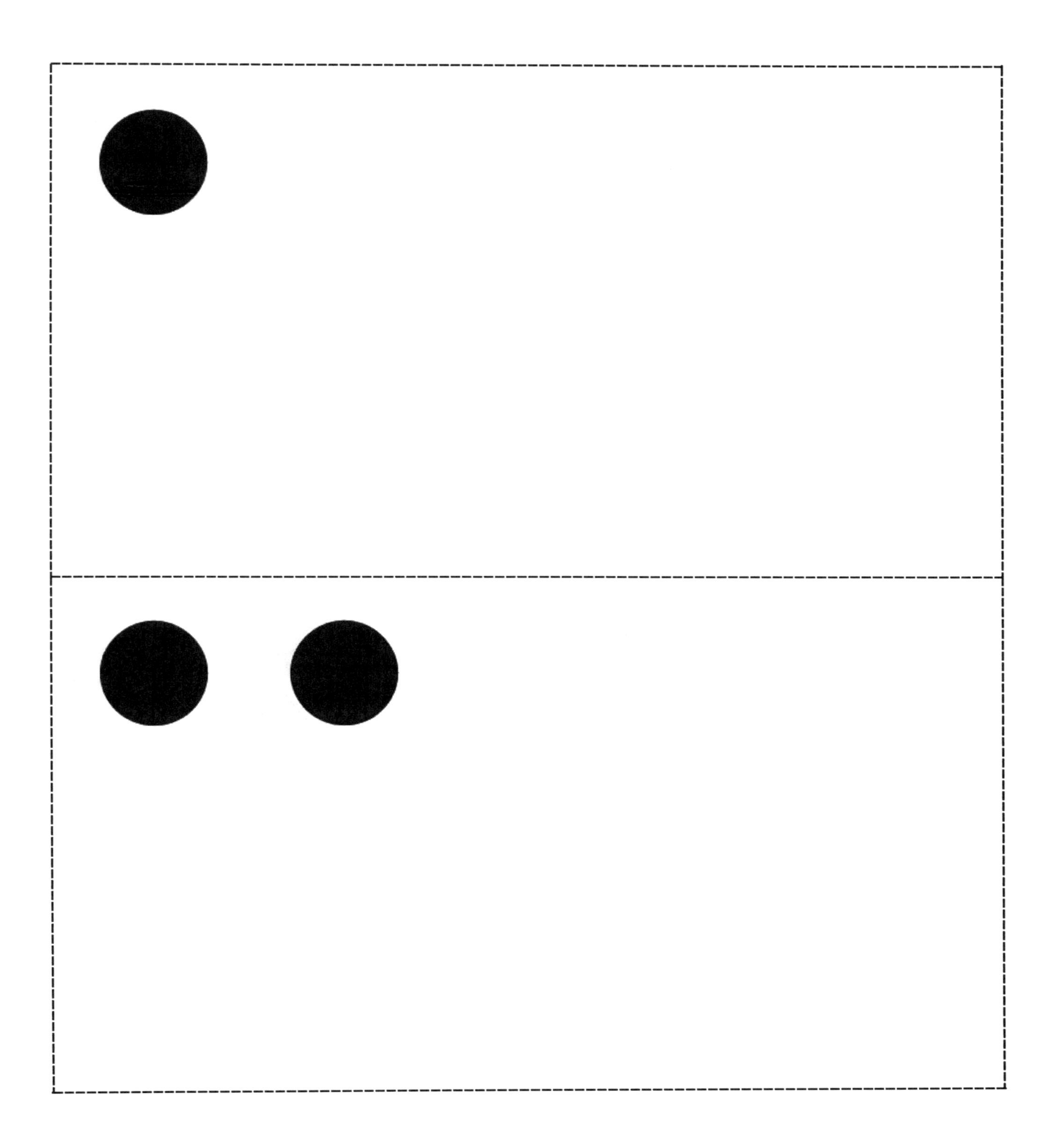

large 5-group cards (Copy on card stock, and cut. Save full set.)

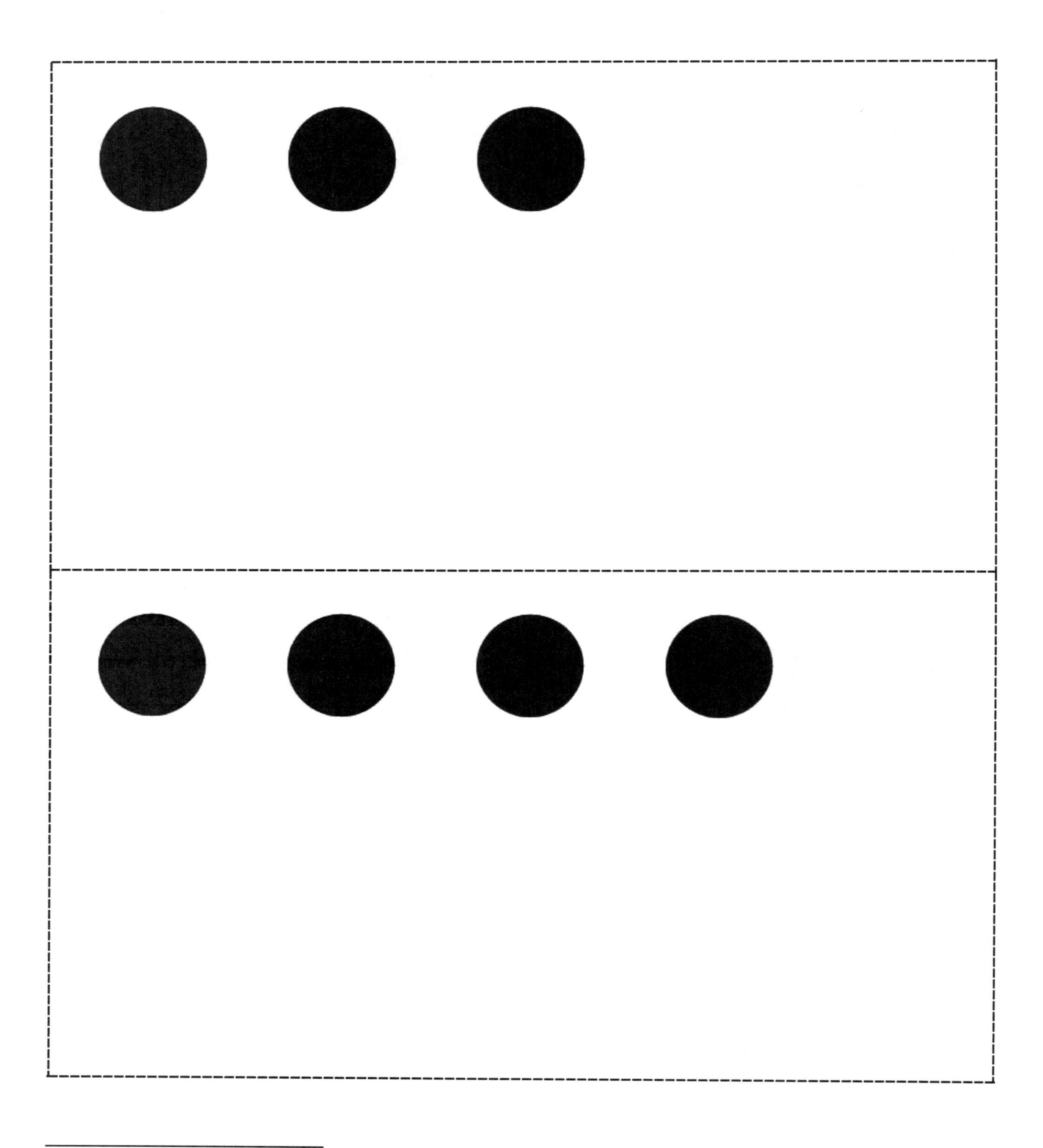

large 5-group cards (Copy on card stock, and cut. Save full set.)

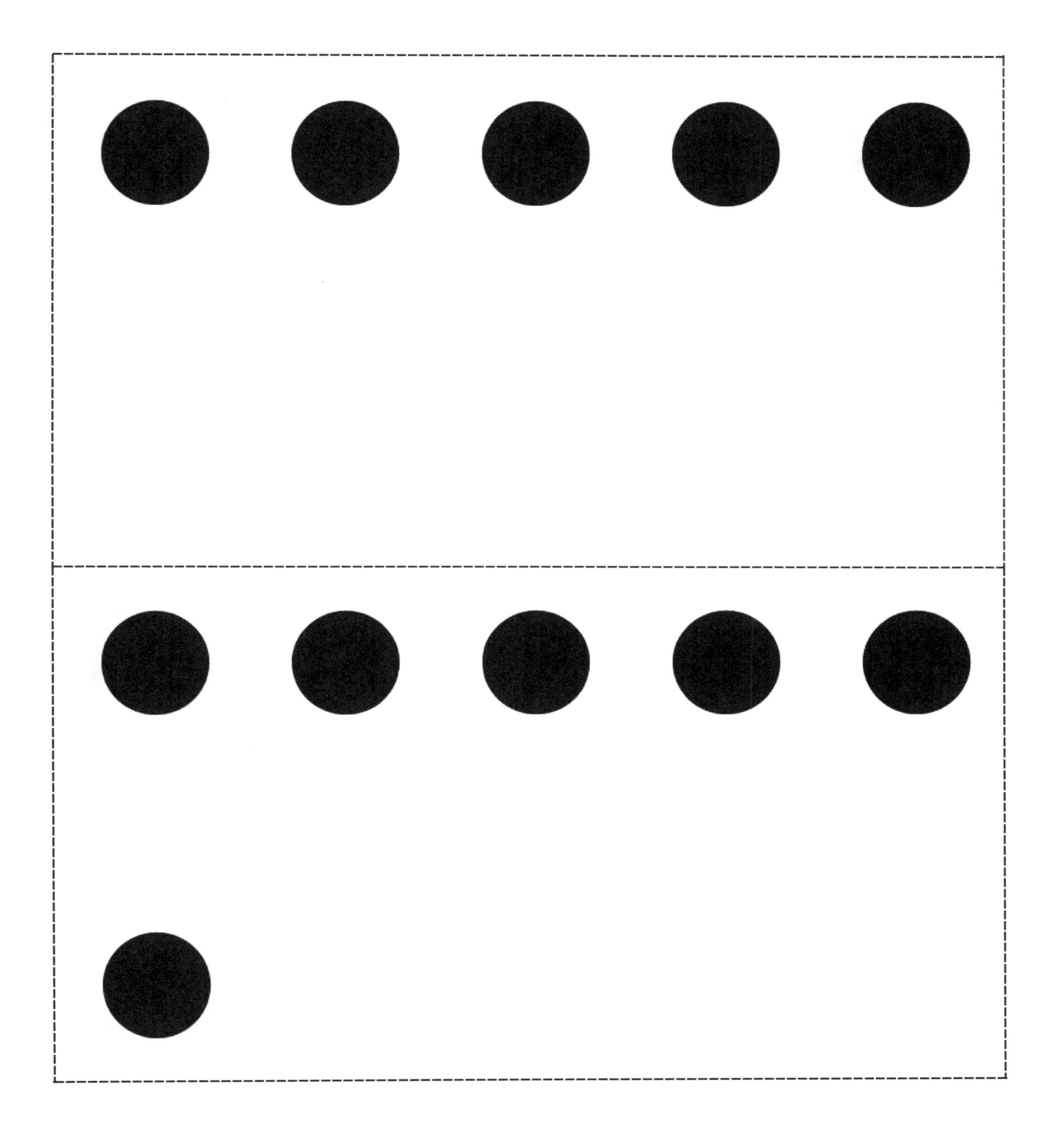

large 5-group cards (Copy on card stock, and cut. Save full set.)

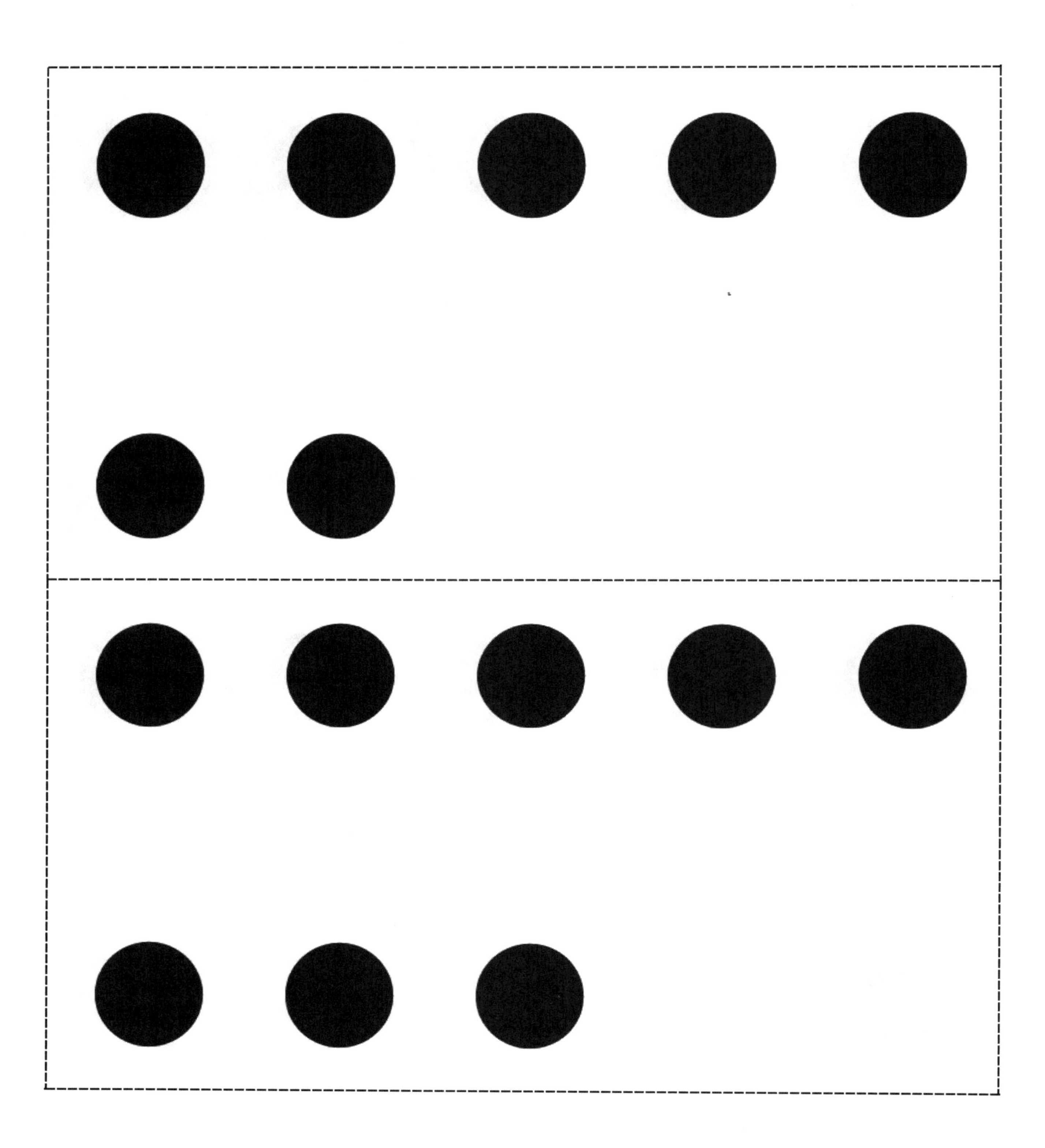

large 5-group cards (Copy on card stock, and cut. Save full set.)

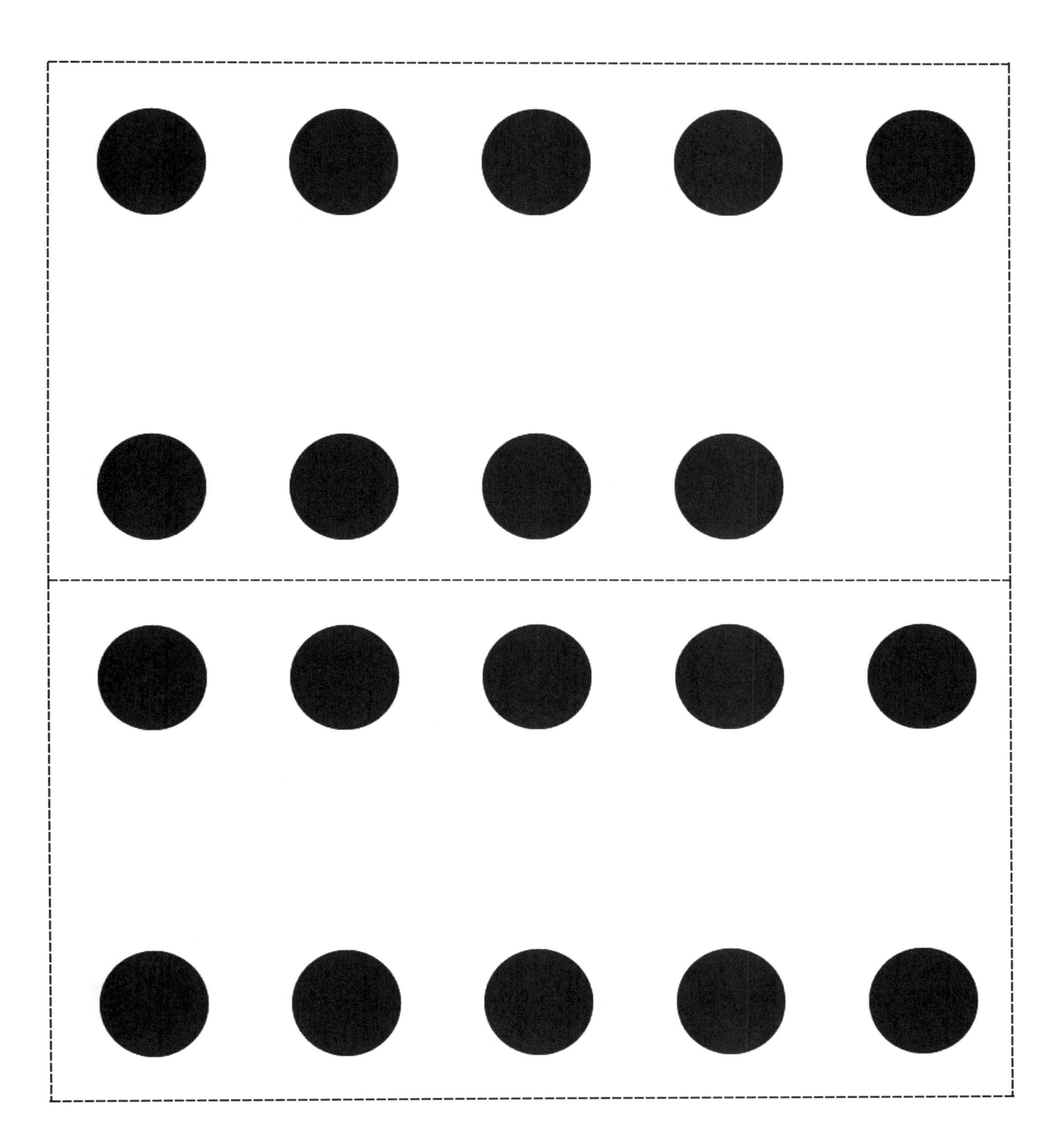

large 5-group cards (Copy on card stock, and cut. Save full set.)

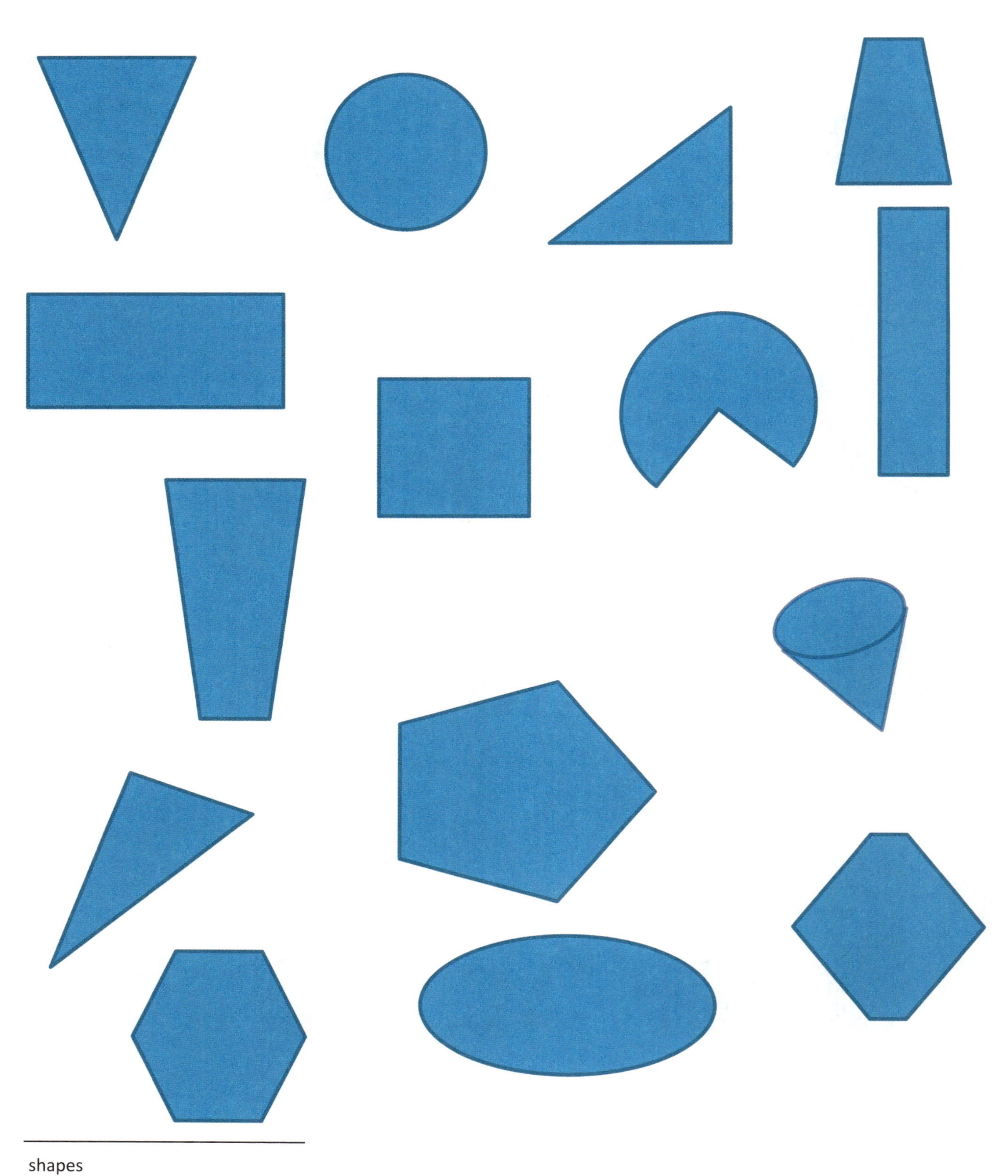

shapes

Lesson 2

Objective: Explain decisions about classifications of triangles into categories using variants and non-examples. Identify shapes as triangles.

Suggested Lesson Structure

Fluency Practice	(12 minutes)
Application Problem	(5 minutes)
Concept Development	(25 minutes)
Student Debrief	(8 minutes)
Total Time	**(50 minutes)**

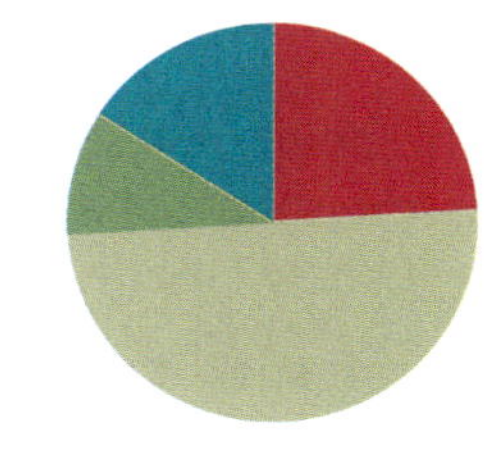

Fluency Practice (12 minutes)

- Making 3 with Triangles **K.OA.3** (3 minutes)
- Make a Shape **K.G.4** (5 minutes)
- Groups of 6 **K.CC.4b** (4 minutes)

Making 3 with Triangles (3 minutes)

Materials: (S) 3 beans, 1 paper or foam triangle, personal white board

Note: This activity was chosen to set the stage for the analysis of triangles coming in today's lesson.

T: Touch and count the corners of the shape.
S: 1, 2, 3.
T: Touch and count your beans.
S: 1, 2, 3.
T: Our job is to make 3. Put 2 beans on the corners of your shape. Keep the other bean in your hand. How many beans are on your shape?
S: 2.
T: How many beans are in your hand?
S: 1.
T: We can tell how to make 3 like this: 2 and 1 make 3. Echo me, please.
S: 2 and 1 make 3.
T: Show me 1 bean on your shape. Keep the rest in your hand. How many beans on your shape?
S: 1.
T: How many beans in your hand?
S: 2.

T: Raise your hand when you can say the sentence, and start with 1. (Wait until all hands are raised, and then give the signal.)

S: 1 and 2 make 3.

Guide students as they write the equations on their personal white boards. Challenge students to list and verify that they have found all possible combinations.

Make a Shape (5 minutes)

Materials: (S) Craft sticks or straws of two different lengths, foam or construction paper work mat

Note: Refrain from naming the shapes at this point. Ask students, if they know them, to keep the names of the shapes secret for now. If students name the shapes, have them explain their thinking by describing the shape's attributes using informal language: "I knew I made a triangle because it has three corners."

T: Let's play Make a Shape. Put three craft sticks this size (hold up the longer of the two lengths) on your mat.

T: Move the sticks so they make a shape with three points.

S: (Move the sticks to form a triangle shape.)

T: Touch and count the points.

S: 1, 2, 3.

T: Touch and count the sides.

S: 1, 2, 3.

T: Are there any curved sides?

S: No.

T: Trade in your three long sticks for three short ones, like this (show students an example of the shorter length), and put them on your mat.

T: Move the sticks so they make a new shape with three points.

S: (Move the sticks to form a different triangle shape.)

T: Does your shape still have three points? Three sides? No curved sides? (Pause after each question to allow students time to verify.)

S: (Respond to questions.)

T: Now, put one of your sticks back. Get a stick this size (hold up the longer of the two lengths), and put it on your mat.

S: (Place the longer stick on the mat so there are now two short and one long.)

T: Move the sticks so they make a new shape with three points.

S: (Move the sticks to form a different triangle shape.)

Have students count the points and sides again and verify that there are no curved sides so that they realize that the attributes of the shape are the same, even as the shape takes on a different appearance. Have them carefully rotate their work mats to view the shape from different angles.

Here is a suggested sequence with names of shapes listed for the teacher's reference:

1. A triangle composed of two long sticks and one short
2. A square composed of four long sticks
3. A smaller square composed of four short sticks
4. A rectangle composed of two short sticks and two long sticks

Groups of 6 (4 minutes)

Note: This maintenance fluency activity helps students gain efficiency in counting objects in varied configurations.

T: When the music starts, calmly walk around the room, visiting corners of the room until you and your classmates can make a group of 6. Don't forget to count yourself! How many can be in a group?

S: 6.

T: So, if you go to a corner that already has 5 people there, can you stay?

S: Yes!

T: What if there are already 6?

S: No.

T: Remember to check all the corners of the room. See if we can all get into groups of 6 before the music stops!

If there are not enough students to make equal groups of the designated number, supplement with puppets or stuffed animals. Allow students to share strategies for making groups quickly.

Application Problem (5 minutes)

It's pizza time! On a piece of paper, draw a large, round pizza pie. Don't forget your favorite toppings! With your crayons, show how you would cut the pizza into enough slices for your family. Compare your slices to those of a partner. Are they alike? Carefully describe the shape of a slice to your partner.

Note: The purpose of this problem is two-fold; first, to have the students create three-sided figures, and second, to set up a potential non-example for use later in the lesson. The curved edge of the crust in their drawing means that the slices are not actually triangles.

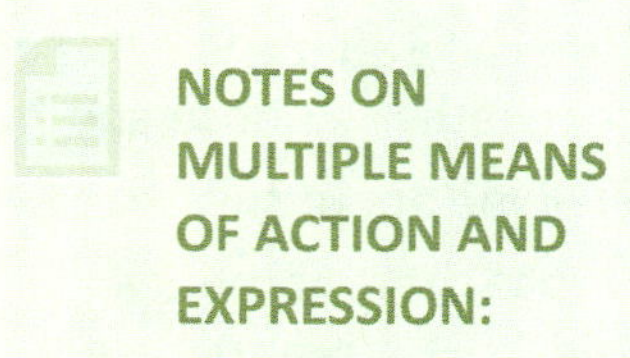

NOTES ON MULTIPLE MEANS OF ACTION AND EXPRESSION:

Scaffold the Application Problem for students who struggle by giving directions one at a time and waiting until students complete the task they were given before giving them the next direction. For example, say, "Draw a large pizza pie." After students comply, continue with, "Use your crayon to cut the pizza into slices for two friends."

Concept Development (25 minutes)

Preparation: Create outlines of geometric figures on paper to be affixed to the board during the lesson (Template). Shapes should include, but not be limited to, those illustrated below:

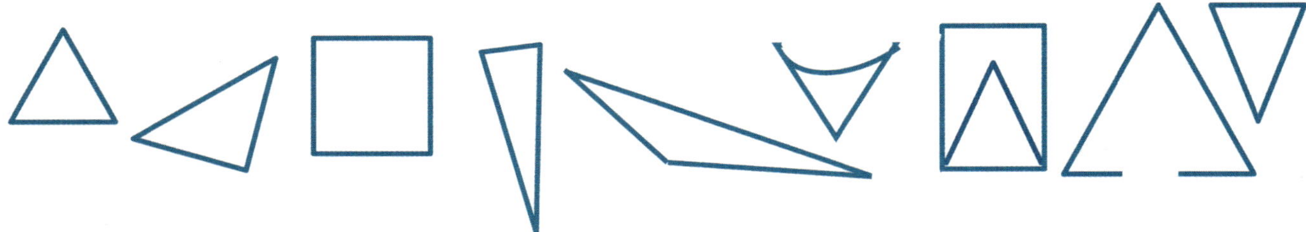

Materials: (S) Geoboard, rubber band

Note: If this is the first time the students have used a **geoboard**, allow a few extra minutes during the lesson to instruct them in proper use of the materials. Using a rubber band, the students stretch the rubber band around pegs to create various shapes. Emphasize that the rubber band must remain on the geoboard at all times.

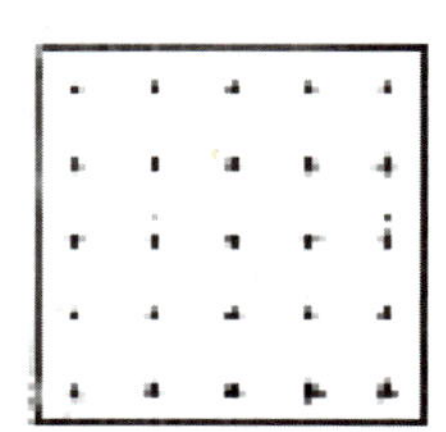

T: Yesterday, when you were telling me about your shapes, you used a lot of math words to describe them. What were some of the things you noticed?

S: Corners. → Curved lines. → Straight. → Number of sides. → Pieces missing.

T: We are going to look at some more shapes today to see what else you notice. (Put a triangle on the classroom board.)

T: Tell me about this shape.

S: It has three sides. → It has three corners. → It doesn't have any curves.

T: We call a shape like this a **triangle**. (Write the word *Triangle* on the board, and affix the shape beneath it. Choose another triangle outline.)

T: Tell me about this shape.

S: It has three corners and three sides. → It has straight sides. → It is a triangle!

T: (Affix the shape to the board under the first triangle.) I am beginning to see a pattern! How many corners does each shape have? (Three.) How many sides? (Three.) What do the sides look like?

S: They are all straight!

T: So, a triangle has three straight sides and three corners?

S: Yes.

T: (Choose 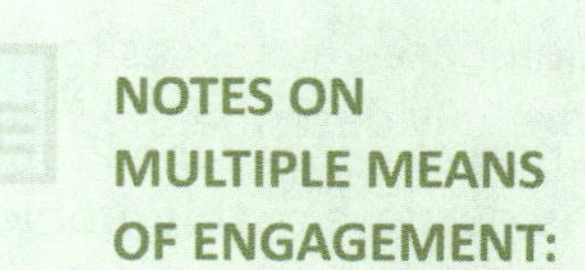.) Here is another shape. It has three corners, and all of the sides are straight. It must be a triangle.

S: No! It's open!

NOTES ON MULTIPLE MEANS OF ENGAGEMENT:

Support English language learners' capacity to discuss how the shapes they made on their geoboards are examples of triangles by providing them with sentence frames such as, "My shape is a triangle because it has..." to use as discussion starters with their partners.

T: If you were a pet inside this fence, you could escape! So, triangles have to be closed?

S: Yes!

T: (Write *Not a Triangle* on the board, and place the shape beneath it. Continue the discussion and sorting with the rest of the shapes, guiding students to point out specific attributes of the variants, non-examples, and distractors.)

T: We have several triangles on the board. I'm going to ask you to copy these triangles onto your geoboard. Remember, you can only use one rubber band. Stretch it around three corners! (Demonstrate, and then pass out geoboards and rubber bands. Assist students as they try to copy the shapes. Make sure that they have shapes with exactly three sides.)

MP.3

T: Now, create your own triangle on your geoboard, and then show your partner. Be sure to tell how you know it is a triangle! (Allow time for sharing and discussion.)

T: Put your geoboards away, and get ready for some triangle hunting on your Problem Set.

Problem Set (10 minutes)

Students should do their personal best to complete the Problem Set within the allotted time.

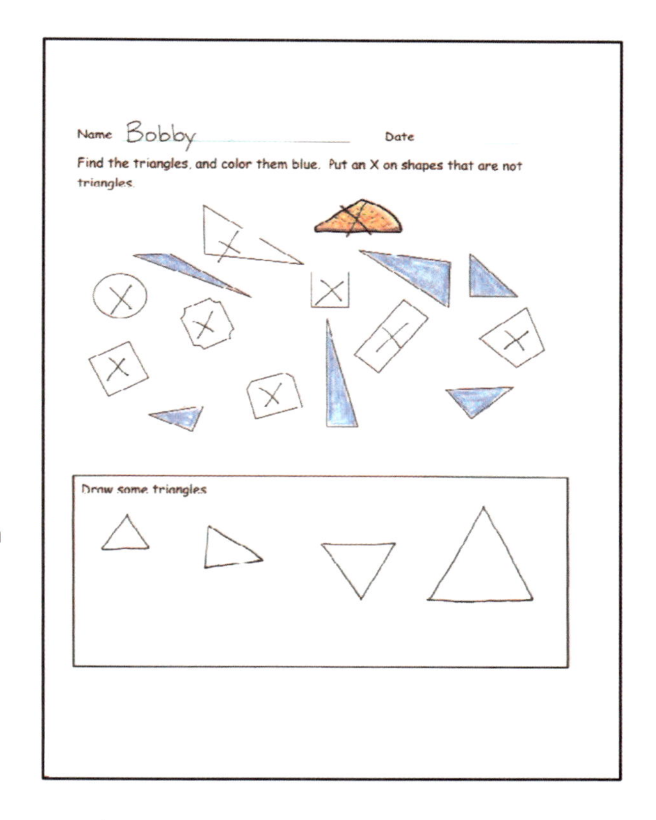

Student Debrief (8 minutes)

Lesson Objective: Explain decisions about classifications of triangles into categories using variants and non-examples. Identify shapes as triangles.

The Student Debrief is intended to invite reflection and active processing of the total lesson experience.

Invite students to review their solutions for the Problem Set. They should check work by comparing answers with a partner before going over answers as a class. Look for misconceptions or misunderstandings that can be addressed in the Debrief. Guide students in a conversation to debrief the Problem Set and process the lesson.

Any combination of the questions below may be used to lead the discussion.

- What new (or significant) math vocabulary did we use today to communicate precisely?
- Count how many **triangles** you colored. Did your partner color that same number?
- Did you color the same triangles as your partner?
- Explain to your partner how you knew the objects you colored were triangles.
- What do you look for in a triangle?
- Were the slices of the pizza in the Application Problem triangles? Why or why not?

Name ______________________________ Date ______________

Find the triangles, and color them blue. Put an X on shapes that are not triangles.

Draw some triangles.

Name ______________________ Date ____________

Color the triangles red and the other shapes blue.

Draw 2 different triangles of your own.

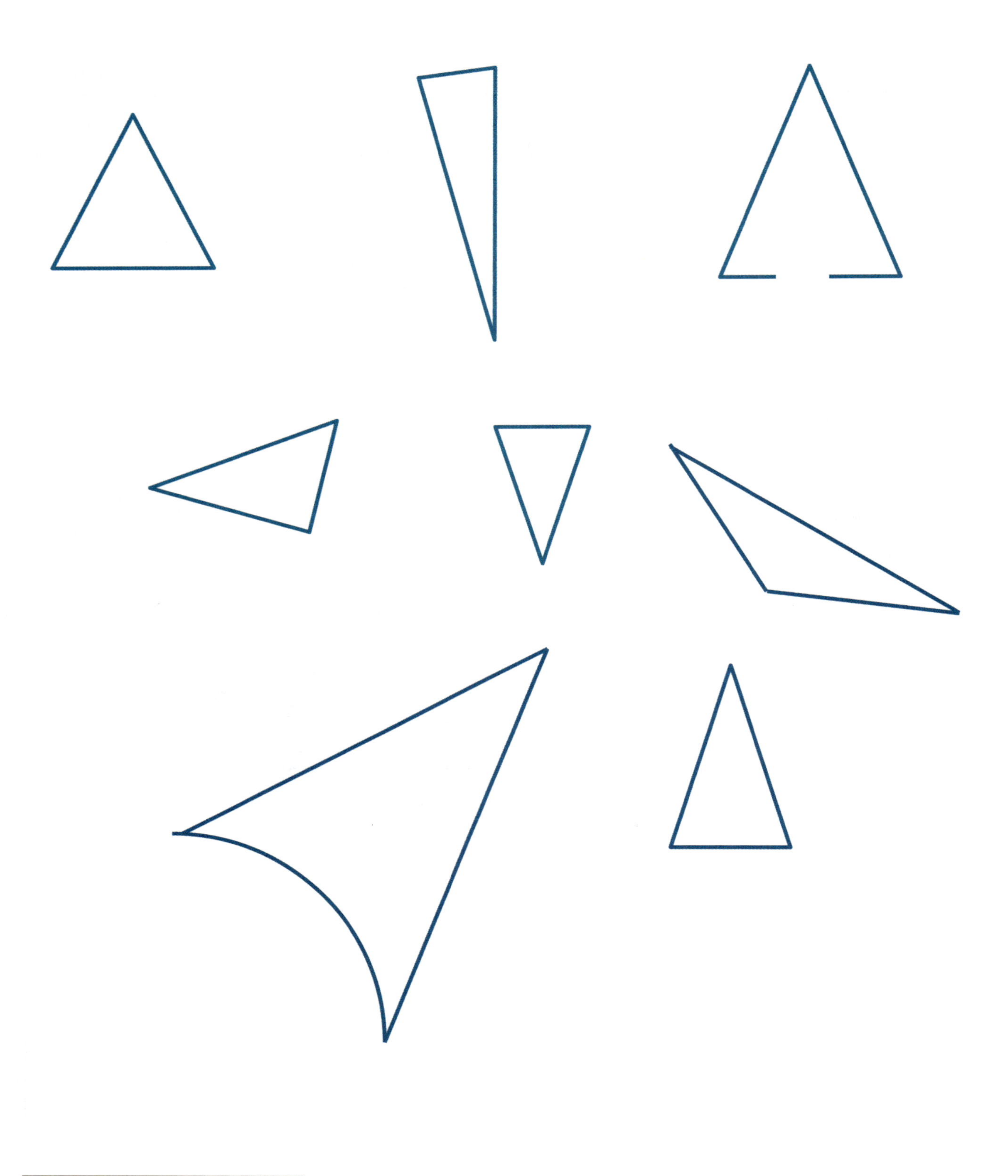

shapes

Lesson 2: Explain decisions about classifications of triangles into categories using variants and non examples. Identify shapes as triangles.

Lesson 3

Objective: Explain decisions about classifications of rectangles into categories using variants and non-examples. Identify shapes as rectangles.

Suggested Lesson Structure

Fluency Practice	(12 minutes)
Application Problem	(5 minutes)
Concept Development	(25 minutes)
Student Debrief	(8 minutes)
Total Time	**(50 minutes)**

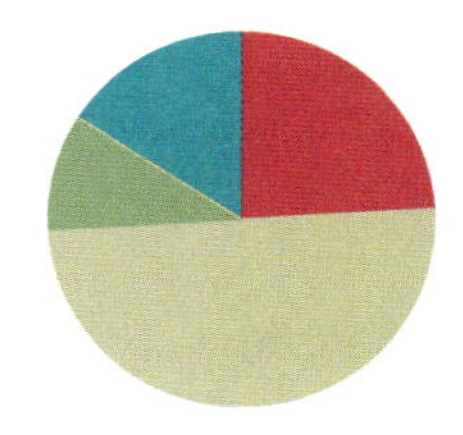

Fluency Practice (12 minutes)

- 5-Group Hands **K.CC.2** (5 minutes)
- Making 4 with Squares and Beans **K.CC.4a** (4 minutes)
- Triangle or Not **K.G.2** (3 minutes)

5-Group Hands (5 minutes)

Materials: (T) Large 5-group cards (Lesson 1 Fluency Template 3)

Note: We repeat work with the hands often because students need frequent practice to achieve fluency. The same exercises must be repeated again and again. As they gain depth of understanding, they visualize. As they visualize, they no longer need their fingers.

Conduct the activity as outlined in Lesson 1, but now continue to 10.

Making 4 with Squares and Beans (4 minutes)

Materials: (S) 4 beans, paper or foam squares, personal white board

Note: Students work early in the year toward fluency with sums and differences within 5. This takes time and a great deal of practice.

T: Touch and count the corners of the square.
S: 1, 2, 3, 4.
T: Touch and count your beans.
S: 1, 2, 3, 4.

T: Our job is to make 4. Put 3 beans on the corners of your square. Keep the other bean in your hand. How many beans on your square?

S: 3.

T: How many beans in your hand?

S: 1.

T: We can tell how to make 4 like this: 3 and 1 make 4. Echo me, please.

S: 3 and 1 make 4.

T: Show me 2 beans on your square. Keep the rest in your hand. How many beans on your square?

S: 2.

T: How many beans in your hand?

S: 2.

T: Raise your hand when you can say the sentence. (Wait until all hands are raised, and then give the signal.)

S: 2 and 2 make 4.

Continue with placing 1 bean on the square, then 4, and finally 0 to work through all of the number combinations. Have students write the equations on their personal white boards. Challenge students to list and verify that they have found all possible combinations.

Triangle or Not (3 minutes)

Materials: (T) Paper shapes of the same color in varying sizes, a wide range of exemplars, non-examples, and variants (Fluency Template)

Note: This is a preparatory fluency activity intended to review the previous lesson's work with triangles and prepare students to name and identify rectangles in a similar manner.

T: I'll show you a shape. We'll try to decide if it's a triangle or not. If you think it's a triangle, give me a thumbs-up. If it's not a triangle, thumbs-down. Either way, be ready to explain your choice! Here we go. (Show an exemplar triangle.)

S: (Show thumbs up sign.)

T: You're right! It is a triangle. Who can tell us why?

S: (Give varied responses. Justify with informal language and attributes of the shape.)

Continue identifying shapes as triangles or not triangles. Proceed from simple to complex by starting with the exemplar of each shape, then the non-examples, and then the variants.

Application Problem (5 minutes)

Design your own dollar bill! Draw your dollar bill on a piece of paper. Whose picture will you put in the center? Compare your dollar with your partner's. Tell him about the shape of your bill. How are your dollars alike?

Note: In this problem, the students should, from their own general knowledge, draw some sort of a rectangle. Describing their dollars to their friends leads them to articulate what they already know about the shape in anticipation of today's lesson.

Concept Development (25 minutes)

Preparation: Create outlines of geometric figures on paper to be affixed to the board during the lesson (Template 1). Shapes should include, but not be limited to, those illustrated below:

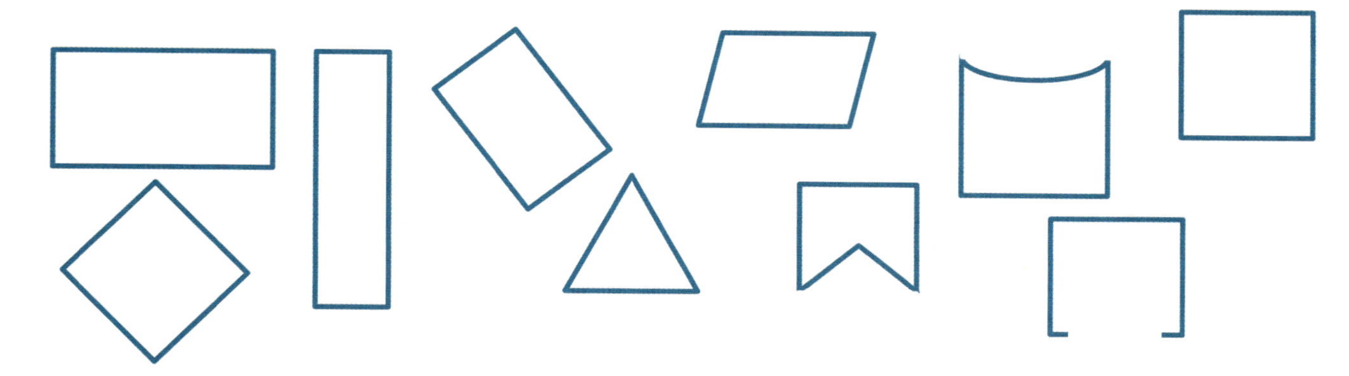

Materials: (S) Copy of dot paper (Template 2) on cardstock, Wikki Stix (crayons or markers may also be used)

T: We are going to talk about another type of shape today. Look at the shape on the board. Use your math words to tell about it. (Place an exemplar **rectangle** on the board.)

S: It has four corners. → It has four sides. → The sides are all straight.

T: This shape is called a **rectangle**. (Write *Rectangle* on board, and affix the shape beneath it.)

T: (Place another rectangle on the board.) Tell about this shape.

S: It has four straight sides and four corners. It is a rectangle, too.

T: Hmmm. I wonder if we will have another pattern today. Let's put this over by the other rectangle. How about this shape? (Place a square on the board.)

S: It has four corners and four sides. All the sides look the same.

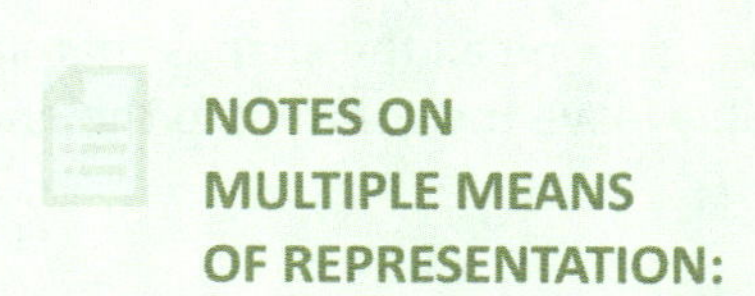

NOTES ON MULTIPLE MEANS OF REPRESENTATION:

For English language learners, connect the key vocabulary of the lesson (*straight, sides, corners*) with pictures that represent the meaning of these terms.

T: So, this is a rectangle, too? (Yes.) This special rectangle, in which all the sides are the same length, is called a…

S: **Square**!

T: How about this one? (Affix ▱ to board.) Is this a rectangle? It has straight sides and four corners.

S: No! It's not a rectangle! → The corners don't look right.

T: What do the corners look like in a rectangle?

S: They need to be L-shaped.

T: Let's put this over here, then. It is not a rectangle. (Write *Not a Rectangle* on the board, and affix the shape beneath it.)

As you did with the triangles yesterday, continue to sort the rest of the shapes with the students. Be sure to place the shapes in a variety of orientations. Guide them to point out pertinent attributes of variants, distractors, and non-examples. Encourage them to insist that any rectangles have four straight sides and four right angles and are closed shapes.

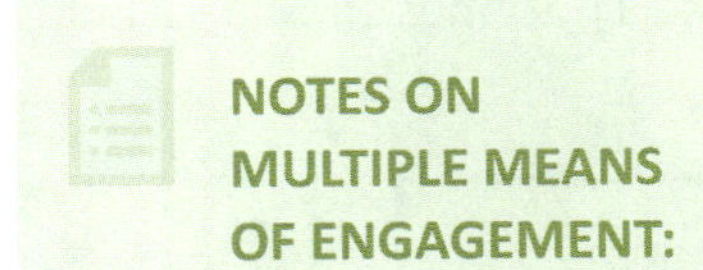

NOTES ON MULTIPLE MEANS OF ENGAGEMENT:

Challenge your students working above grade level by asking them to squeeze their Wikki Stix rectangles (making a parallelogram), and have them decide whether the new shape is a rectangle or not.

T: You have learned many rules about rectangles today! Now, make some rectangles of your own. Use these Wikki Stix for the sides, and use the special dots on this paper to keep your sides straight. Use the rectangles you sorted on the board for your models.

Pass out materials, and allow time for construction and discussion. Circulate to ensure accuracy. If Wikki Stix are not available, students can use crayons or markers to connect the dots in their shape construction.

T: Great work! When you take these home after school, see if you notice any rectangles like these on the way!

Problem Set (8 minutes)

Students should do their personal best to complete the Problem Set within the allotted time.

In this Problem Set, we suggest all students begin by putting an X on all the shapes that are not rectangles and possibly leave the coloring to the end if they still have time.

Name Aly Date 10-3-13

Find the rectangles and color them red. Put an X on shapes that are not rectangles.

Draw rectangles.

Student Debrief (8 minutes)

Lesson Objective: Explain decisions about classifications of rectangles into categories using variants and non-examples. Identify shapes as rectangles.

The Student Debrief is intended to invite reflection and active processing of the total lesson experience.

Invite students to review their solutions for the Problem Set. They should check work by comparing answers with a partner before going over answers as a class. Look for misconceptions or misunderstandings that can be addressed in the Debrief. Guide students in a conversation to debrief the Problem Set and process the lesson.

Any combination of the questions below may be used to lead the discussion.

- How did the Application Problem connect to today's lesson?
- What new (or significant) math vocabulary did we use today to communicate precisely?
- Count how many **rectangles** you colored. Did your partner color that same number?
- Did you color the same rectangles as your partner?
- Explain to your partner how you knew the objects you colored were rectangles.
- What do you look for in a rectangle?
- What shape did you draw with four sides? Can you draw more than one shape with four sides?
- How are rectangles and triangles the same and different?
- Why is a **square** a special kind of rectangle?

Name ______________________________ Date ______________

Find the rectangles, and color them red. Put an X on shapes that are not rectangles.

Draw some rectangles.

Name ______________________________ Date ______________

Color all the rectangles red. Color all the triangles green.

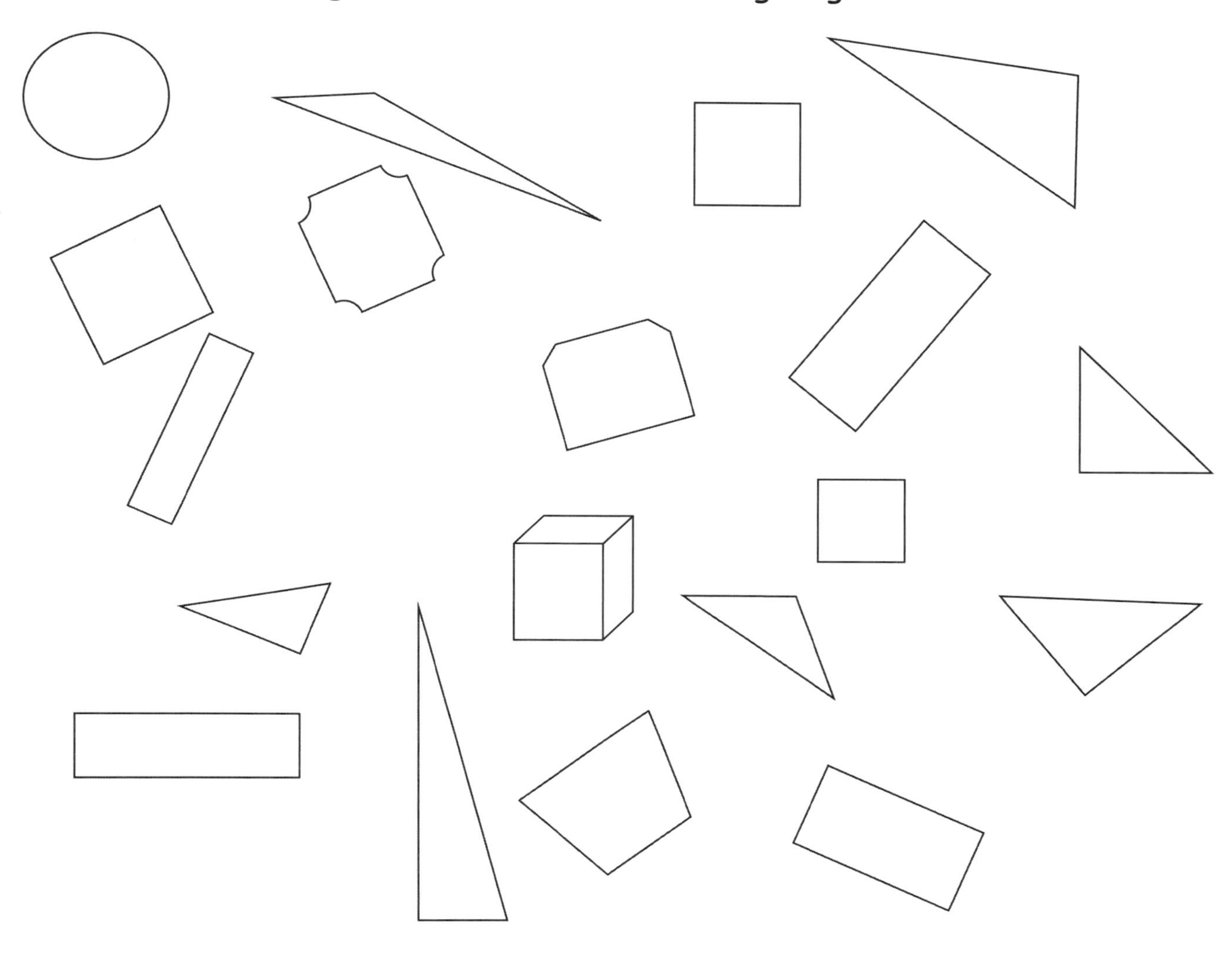

On the back of your paper, draw 2 rectangles and 3 triangles. How many shapes did you draw? Put your answer in the circle.

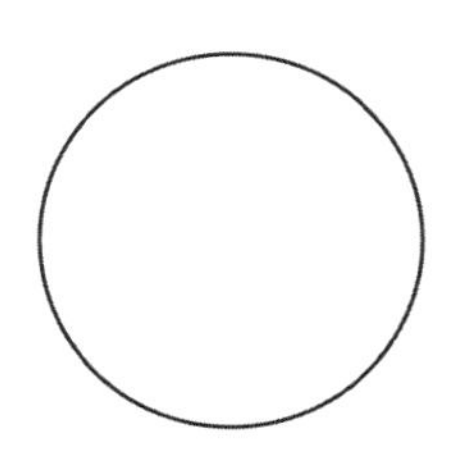

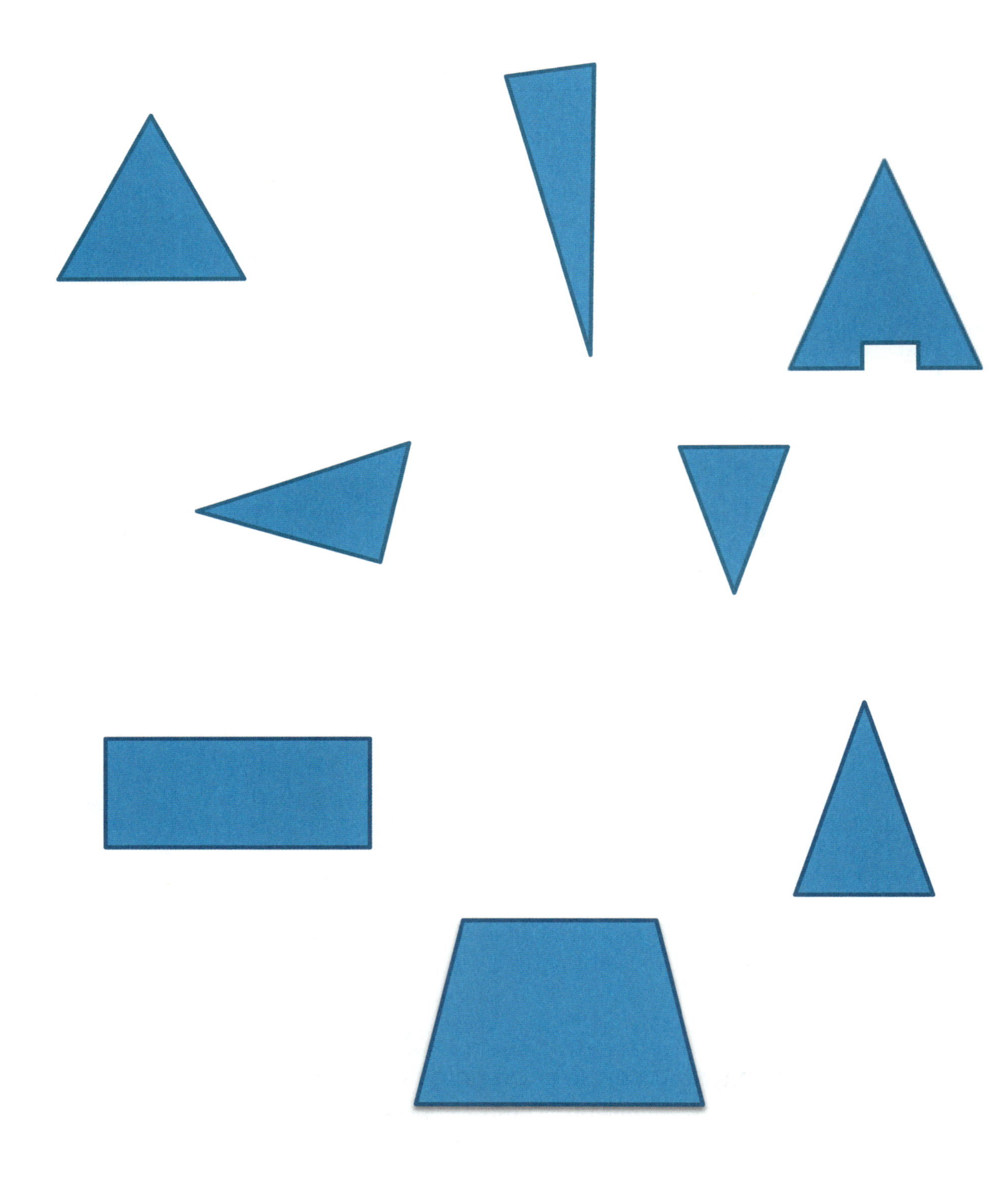

shapes

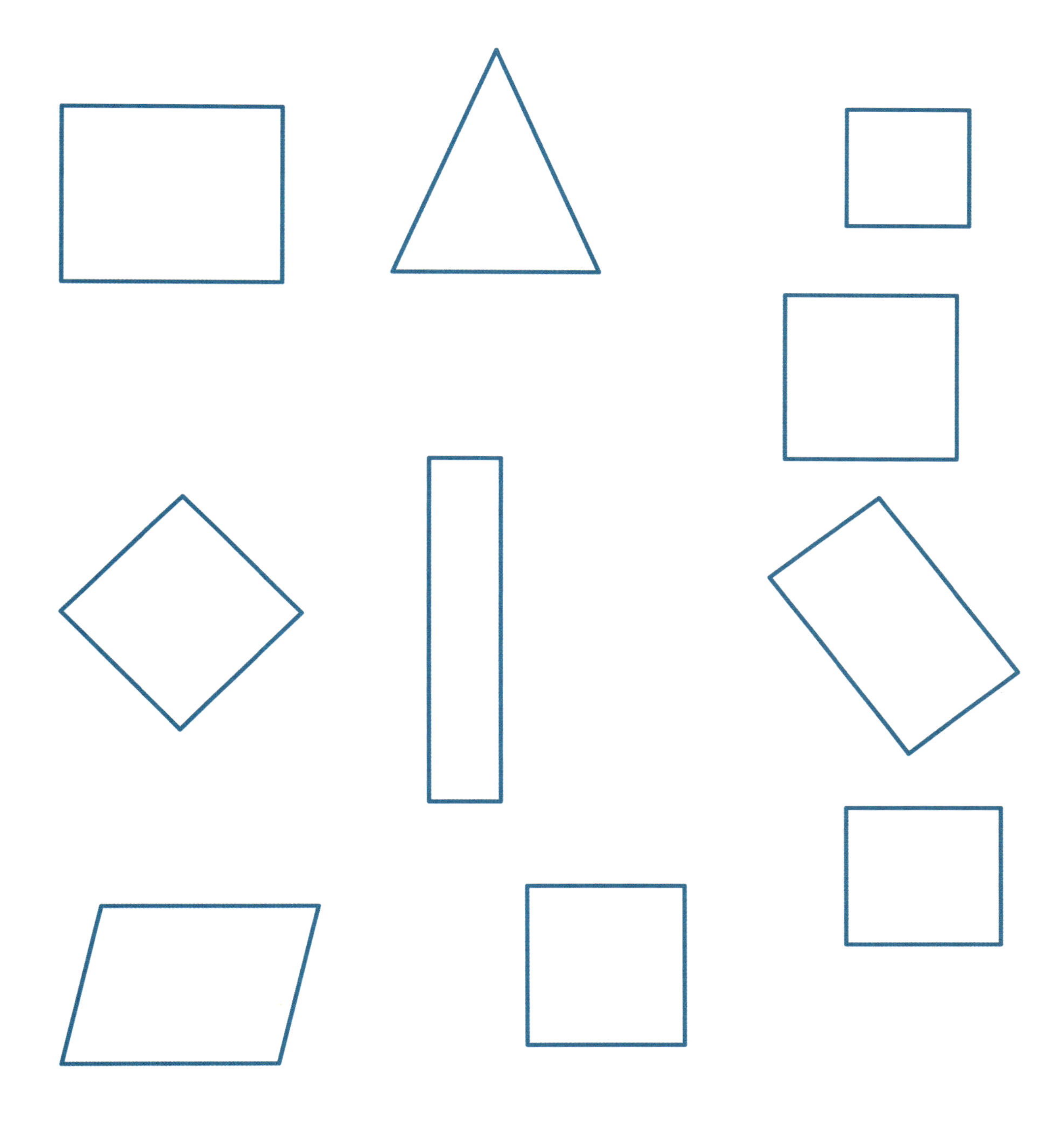

shapes

dot paper

Lesson 4

Objective: Explain decisions about classifications of hexagons and circles, and identify them by name. Make observations using variants and non-examples.

Suggested Lesson Structure

- Fluency Practice (12 minutes)
- Application Problem (5 minutes)
- Concept Development (25 minutes)
- Student Debrief (8 minutes)

Total Time **(50 minutes)**

Fluency Practice (12 minutes)

- Rectangle or Not **K.G.2** (3 minutes)
- Make a Shape **K.G.4** (4 minutes)
- Groups of 7 **K.CC.4b** (5 minutes)

Rectangle or Not (3 minutes)

Materials: (T) Paper shapes of the same color in varying sizes, a wide range of exemplars, non-examples, and variants (Fluency Template)

Note: This is a preparatory fluency activity intended to review the previous lesson's work with rectangles and prepare students to name and identify hexagons and circles in a similar manner.

This is similar to Lesson 3, but with rectangles.

Identify shapes as rectangles or not rectangles, from simple to complex, by starting with the exemplar of each shape, then the non-examples, and then the variants.

Make a Shape (4 minutes)

Note: This activity is repeated with a new shape, allowing students to focus on the new component, the hexagon, rather than the logistics of the activity itself.

Conduct the activity as outlined in Lesson 2, but this time include hexagons without naming.

Groups of 7 (5 minutes)

Note: This maintenance activity supports efficiency in counting objects in varied configurations.

Conduct the activity as outlined in Lesson 2, but with 7. Allow students to share their strategies for making groups quickly.

Application Problem (5 minutes)

Using only triangles and rectangles, design a rocket ship on your paper. Trade rocket ships with your partner. Count how many triangles and rectangles you see in his picture. Did you use the same number of each shape?

Note: This problem is designed as a review exercise prior to the introduction and definition of two new shapes in today's lesson.

Concept Development (25 minutes)

Preparation: While many objects in classrooms have a circular shape, hexagons in the classroom environment usually must be engineered. Strategically place several cutout or outlined shapes of regular and irregular hexagons around the room prior to the lesson. You may wish to include a few different hexagons constructed on geoboards or on dot paper.

Create outlines of geometric figures on paper to be affixed to the board during the lesson. Shapes should include, but not be limited to, those illustrated below:

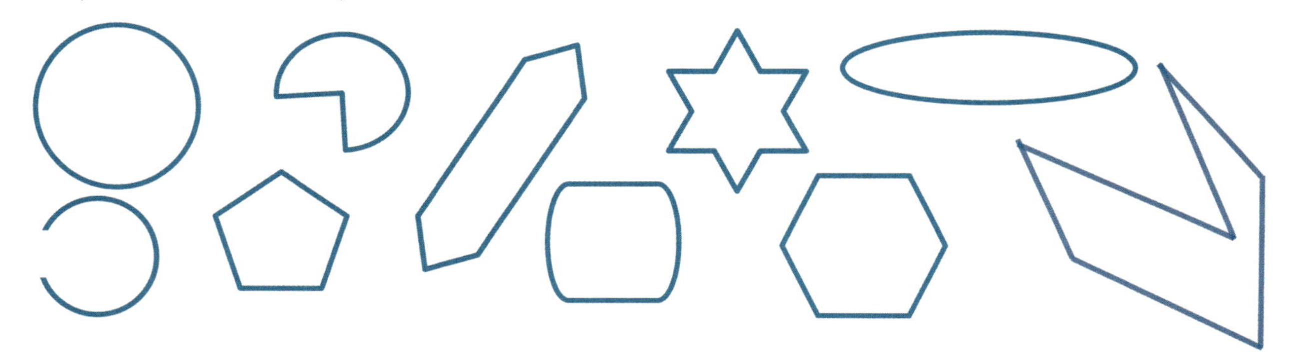

Materials: (T) Paper shapes (Template) (S) Clipboard with paper and pencil, real or toy magnifying glass (if available)

> T: We have talked about triangles, rectangles, and squares, and you have made many of these yourselves. Here are some new shapes for you to look at today.

Sort shape outlines to make a group of **hexagons** and then **circles** in the same manner as in Lessons 2 and 3. As in previous lessons, begin the discussion with exemplar shapes to guide the students as they discover each shape's defining characteristics. As sorting continues, guide them to realize that a hexagon is a closed flat shape with six straight sides and that a circle is a flat, closed, curved shape with no straight sides.

Note: Students can become frustrated as they attempt to articulate the difference between a circle and an oval. Though they may not be able to describe the concept of equidistance from a center, they can tell you that if they had a race car, they would rather have wheels in the shape of a circle than in the shape of an oval. "Circles can roll better!" "They are not squished!"

MP.1

T: We are going to have another detective hunt today. You and your partner will search for these shapes in the classroom. Use your clipboards and detective equipment, and draw any circles and hexagons that are hiding! (Allow students to investigate for five minutes before they return to their seats.)

T: Would anyone like to show and share one of the circles or hexagons they found in the classroom today? How is your circle or hexagon different from the other shapes we've learned? (Allow time for sharing and discussion.)

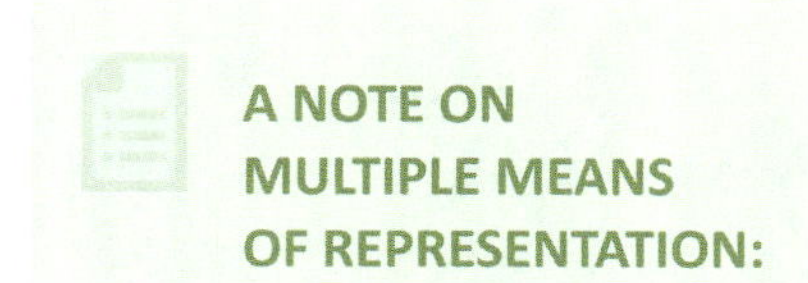

A NOTE ON MULTIPLE MEANS OF REPRESENTATION:

Once the vocabulary words *hexagon* and *circle* have been introduced, post these on the word wall with a visual of a circle and many different examples of hexagons.

Problem Set (10 minutes)

Students should do their personal best to complete the Problem Set within the allotted time.

Since hexagons and circles are the focus of this lesson, have students first identify one of the two shapes and then the other.

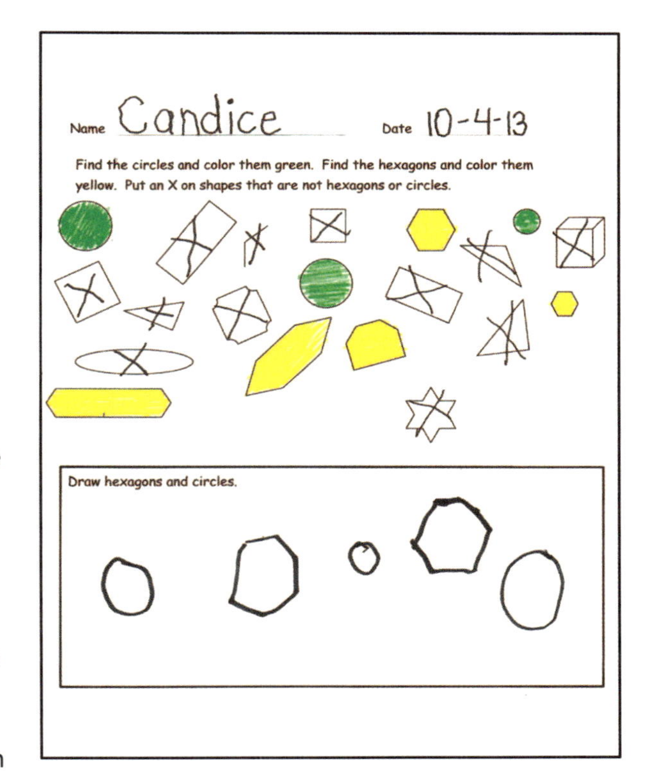

Student Debrief (8 minutes)

Lesson Objective: Explain decisions about classifications of hexagons and circles, and identify them by name. Make observations using variants and non-examples.

The Student Debrief is intended to invite reflection and active processing of the total lesson experience.

Invite students to review their solutions for the Problem Set. They should check work by comparing answers with a partner before going over answers as a class. Look for misconceptions or misunderstandings that can be addressed in the Debrief. Guide students in a conversation to debrief the Problem Set and process the lesson.

Any combination of the questions below may be used to lead the discussion

- How did the Application Problem connect to today's lesson?
- What new (or significant) math vocabulary did we use today to communicate precisely?
- Did you color the same **hexagons** and **circles** as your partner?
- Explain to your partner how you knew the objects you colored were hexagons or circles.

- Count how many circles and hexagons you colored. Did your partner color that same number?
- Which shape is more like a circle, a square or a hexagon with equal sides? If there were more and more equal sides to our shape, could it look more and more like a circle?

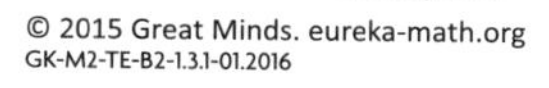

Name ______________________________ Date ______________

Find the circles, and color them green. Find the hexagons, and color them yellow. Put an X on shapes that are not hexagons or circles.

Draw hexagons and circles.

Name ______________________________ Date ______________

Color the triangles blue.
Color the rectangles red.
Color the circles green.
Color the hexagons yellow.

On the back of your paper, draw 2 triangles and 1 hexagon.

How many shapes did you draw? ______

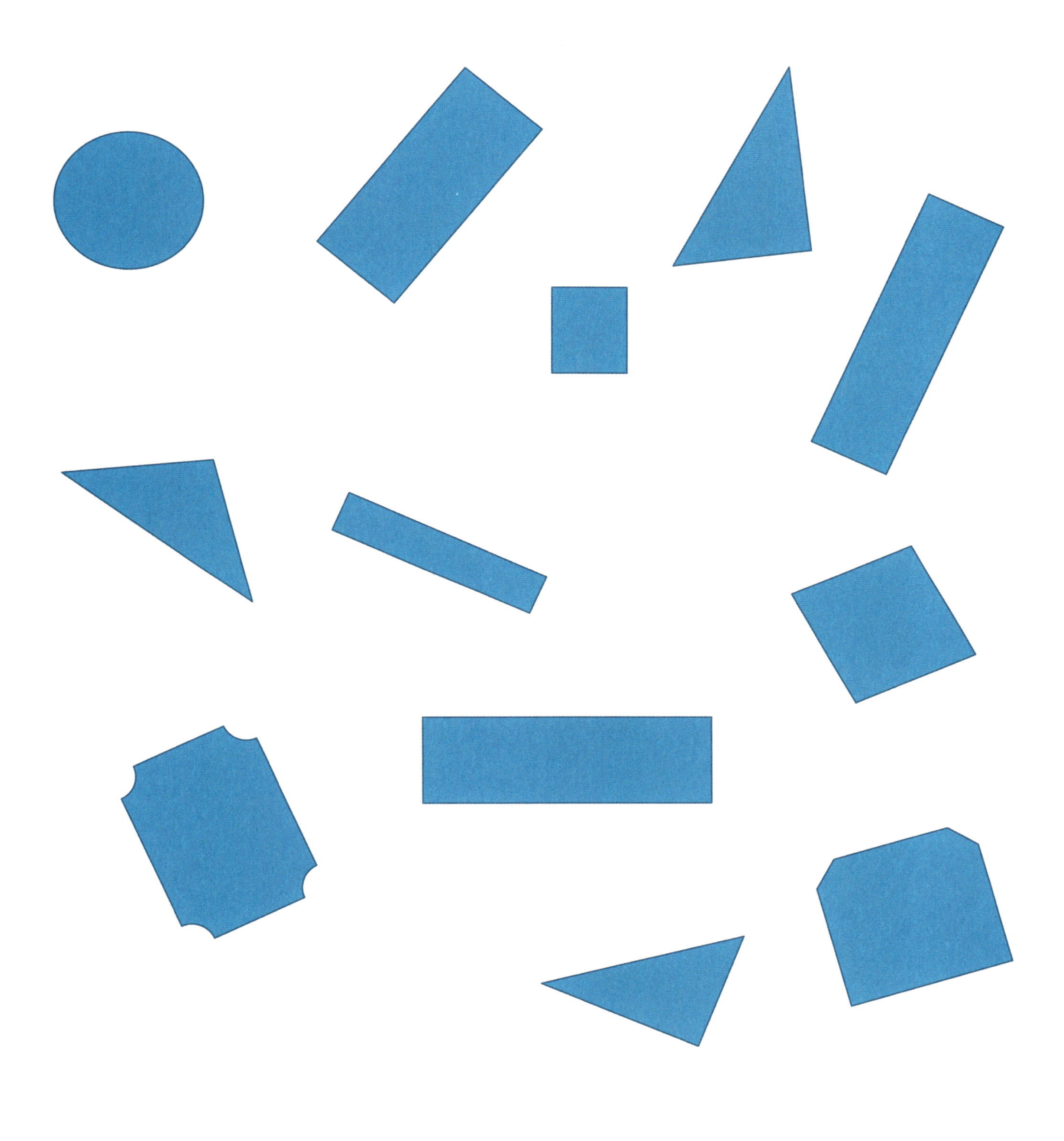

paper shapes

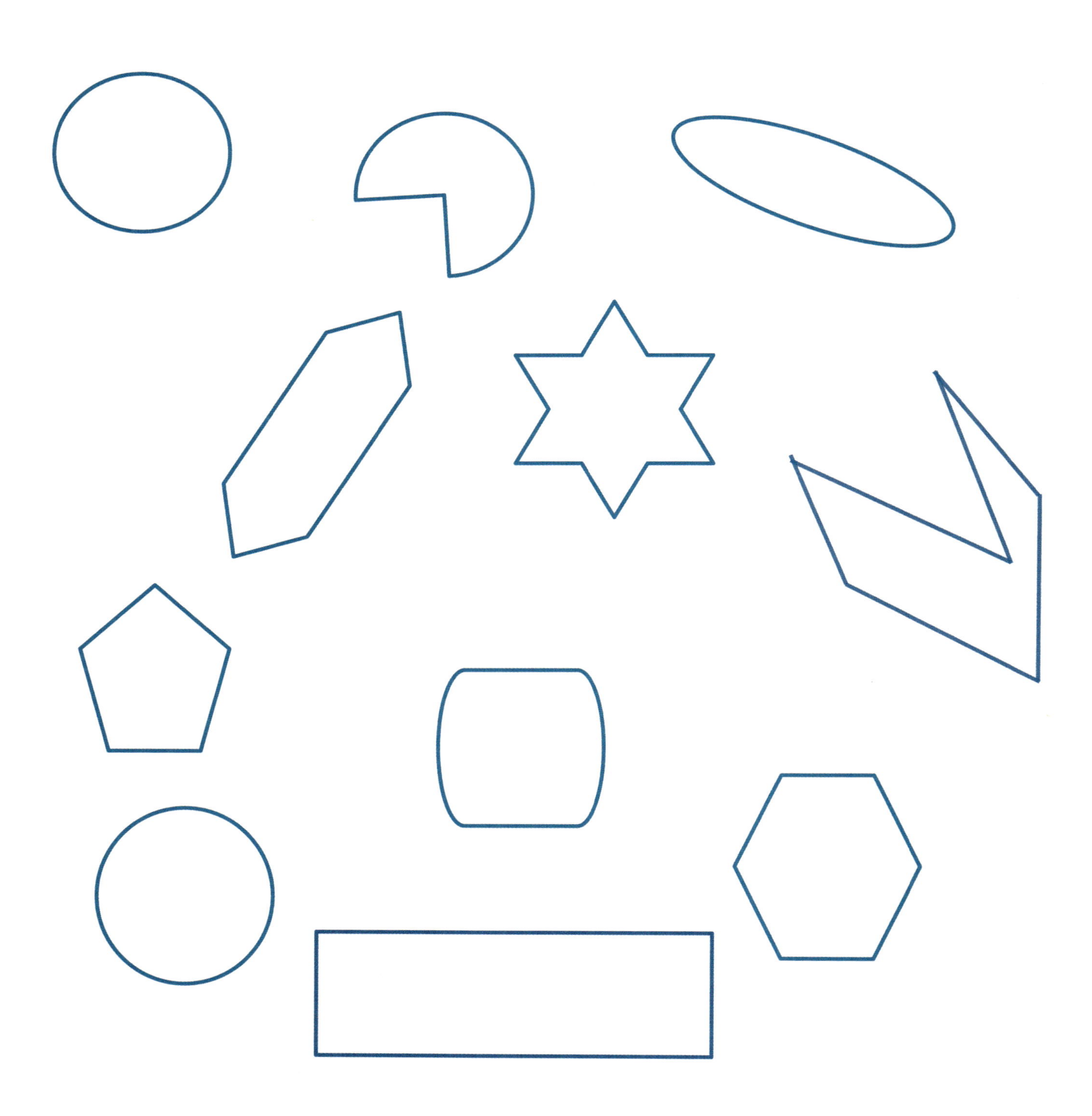

paper shapes

Lesson 5

Objective: Describe and communicate positions of all flat shapes using the words *above, below, beside, in front of, next to*, and *behind*.

Suggested Lesson Structure

Fluency Practice	(12 minutes)
Application Problem	(5 minutes)
Concept Development	(25 minutes)
Student Debrief	(8 minutes)
Total Time	**(50 minutes)**

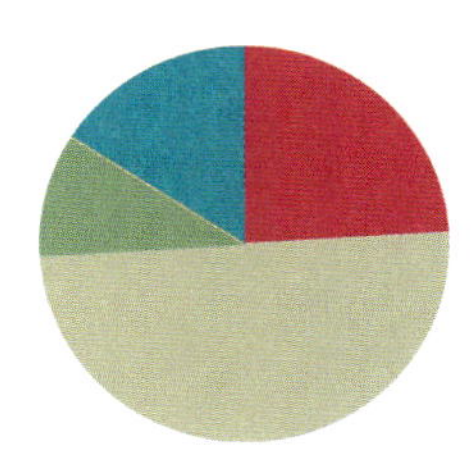

Fluency Practice (12 minutes)

- Groups of Shapes **K.G.2** (5 minutes)
- Peek-a-Boo Shapes **K.G.2** (3 minutes)
- Groups of 8 **K.CC.4b** (4 minutes)

Groups of Shapes (5 minutes)

Materials: (T) Signs with pictures of shapes to indicate where to form each group (Fluency Template 1)
(S) Paper cutouts of triangles, rectangles, squares, hexagons, and circles (variety of sizes, include exemplars, non-examples, and variants) (Fluency Template 2)

Note: In this fluency activity, students bring together their insights from Lessons 1–4 to form groups, each defined by a shape. This allows teachers to also use shapes as part of their classroom management techniques, which will further embed geometry into the classroom culture.

T: Choose a shape, and then meet me at the rug.
T: Look at your shape. Raise your hand if you know the name of your shape. When I give the signal, whisper the name of your shape to yourself. Ready?
T: Look around the room. Do you see signs with pictures of shapes?
S: Yes.
T: Do you see your shape?
S: Yes.
T: When I start the music, I want you to calmly walk to the sign that has the same shape as yours.
T: When I point to your group, say the name of your shape. (Point to the group of triangles.)
S: Triangles.

Continue identifying the remaining groups, and then call students back to the rug to trade for a new shape. Circulate to see which students struggle with this task. Support them by having them identify the attributes of their shape and compare it to the shapes pictured on the signs.

Peek-a-Boo Shapes (3 minutes)

Materials: (T) Paper cutouts of triangles, rectangles, squares, hexagons, and circles (variety of sizes, include exemplars, non-examples, and variants), pictures of real-world objects that are flat shapes (Fluency Template 3)

Note: With the teacher hiding the shapes, students get accustomed to visualizing, a skill they will be applying to numbers, for example, with dot cards. This is an imperative step in developing number sense. It is a significant moment when students realize they can make a mental picture of something.

One shape at a time, show students each shape briefly. Then, take the shape out of view. Remind students beforehand that they are to use the *listen, think, raise your hand, wait for the snap* procedure to name the shape in choral response. Start with easy shapes to build confidence, and then steadily increase the level of difficulty.

Groups of 8 (4 minutes)

Note: This fluency activity helps students gain efficiency in counting objects in varied configurations.

Conduct the activity as outlined in Lesson 2, but with 8. Allow students to share their strategies for making groups quickly.

Application Problem (5 minutes)

Work with your partner. Stand somewhere in the classroom so that you are facing a wall, but your partner is facing the other way. Tell your partner several things that you think are behind you in the room. Have him look to see if you are right. When you are done, switch places with your partner.

Note: *Behind* is a preposition with which most children are very familiar. Introducing newer directional concepts with this familiar word sets the stage for learning in the lesson today.

Concept Development (25 minutes)

Materials: (S) Scissors, glue, paper bag containing cutouts of various shapes (two non-identical shapes of each type, including triangles, rectangles, circles, hexagons, and squares) (Template)

The following are suggestions:

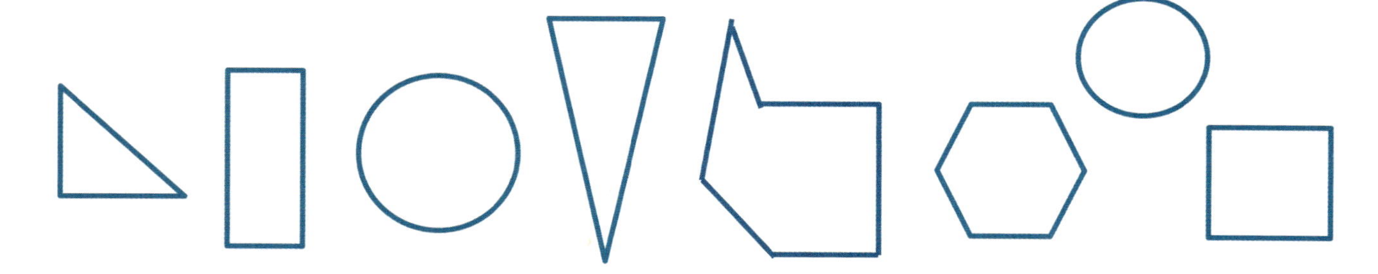

T: You have new mystery bags! Carefully shake your shapes onto your desk. Work with your partner, and say the name of each shape.

S: (Sort out and arrange shapes.)

T: (Allow time for sharing and discussion.) Let's play Simon Says! I will describe a shape to you. If I say "Simon says!" find the shape, hold it up, and freeze. Then, listen carefully while I describe a new shape. Find the new shape only if I say "Simon says!" If I don't, stay frozen.

MP.6

T: Simon says hold up a shape that has four corners. Simon says hold up a shape that has no straight sides. Simon says hold up a shape that has one more than five sides. Hold up a shape that has four sides that are exactly the same.

S: You didn't say "Simon says!"

T: (Continue several times until students show fluency in identifying the correct shapes.)

T: Now, put all of your shapes in a row on the top of your desk. We are going to play our game a different way. I am going to ask you to arrange your shapes in certain ways. Listen carefully, and don't forget to listen for "Simon says!"

T: Simon says put a shape with one less than four sides in the middle of your desk. Simon says put a shape made with a curved line **beside** that shape. Simon says put a shape with only one L-corner **next to** the shape with a curved line. Put a shape with six corners next to that shape. I didn't say "Simon says!" (Play several more times.)

T: Simon says put a curved shape **below** your chair. Simon says hold your square **above** your head. Simon says hold it **behind** your friend. Put a square **in front of** you!

S: You didn't say "Simon says!"

T: Put your shapes in the bag. We are going to practice more words like *below, above*, and *beside* in our Problem Set.

NOTES ON MULTIPLE MEANS OF ACTION AND EXPRESSION:

English language learners are more successful in following Simon Says directions if shown concrete visual examples with the directions. For example, say, "Hold up a shape that has four corners," while pointing to a picture of a corner.

NOTES ON MULTIPLE MEANS OF REPRESENTATION:

Support students who struggle by partnering key words such as *next to, below, above*, and *below* with modeling the actions for them.

Problem Set (10 minutes)

Students should do their personal best to complete the Problem Set within the allotted time.

In this Problem Set, students should begin by cutting out the shapes and lining them next to the duck sheet.

Note: Students will not use all cutout shapes in the Problem Set.

Read the directions for the Problem Set, and then circulate as students work to see if they have mastered the names of the shapes.

- Find the shape with four straight sides that are exactly the same. Color it blue. Glue the shape above the duck.
- Find the shape with no corners. Color it yellow. Glue the shape behind the duck.
- Find the shape with three straight sides. Color it green. Glue the shape below the duck.
- Find the shape with four sides. Two sides are long and the same length, and two sides are short and the same length. Color it red. Glue the shape beside the circle.
- Find the shape with six corners. Color it orange. Glue the shape in front of the duck.
- Find the shape with curves and corners. Color it purple. Glue the shape next to the square.

Student Debrief (8 minutes)

Lesson Objective: Describe and communicate positions of all flat shapes using the words *above, below, beside, in front of, next to*, and *behind*.

The Student Debrief is intended to invite reflection and active processing of the total lesson experience.

Invite students to review their solutions for the Problem Set. They should check work by comparing answers with a partner before going over answers as a class. Look for misconceptions or misunderstandings that can be addressed in the Debrief. Guide students in a conversation to debrief the Problem Set and process the lesson.

Any combination of the questions below may be used to lead the discussion.

- What new (or significant) math vocabulary did we use today to communicate precisely?
- How did you place each object on your paper? Go through each direction (**above, below, in front of, next to**, and **behind**), and compare where students put their objects on their paper.
- Compare with your partner. Did you put your shapes in the same place as your partner?
- What shapes do you see on your paper? How did you know they were those shapes?
- How did the Application Problem connect to today's lesson?

Name ________________________________ Date _______________

Cut out all of the shapes, and put them next to your paper with the duck. Listen to the directions, and glue the objects onto your paper.

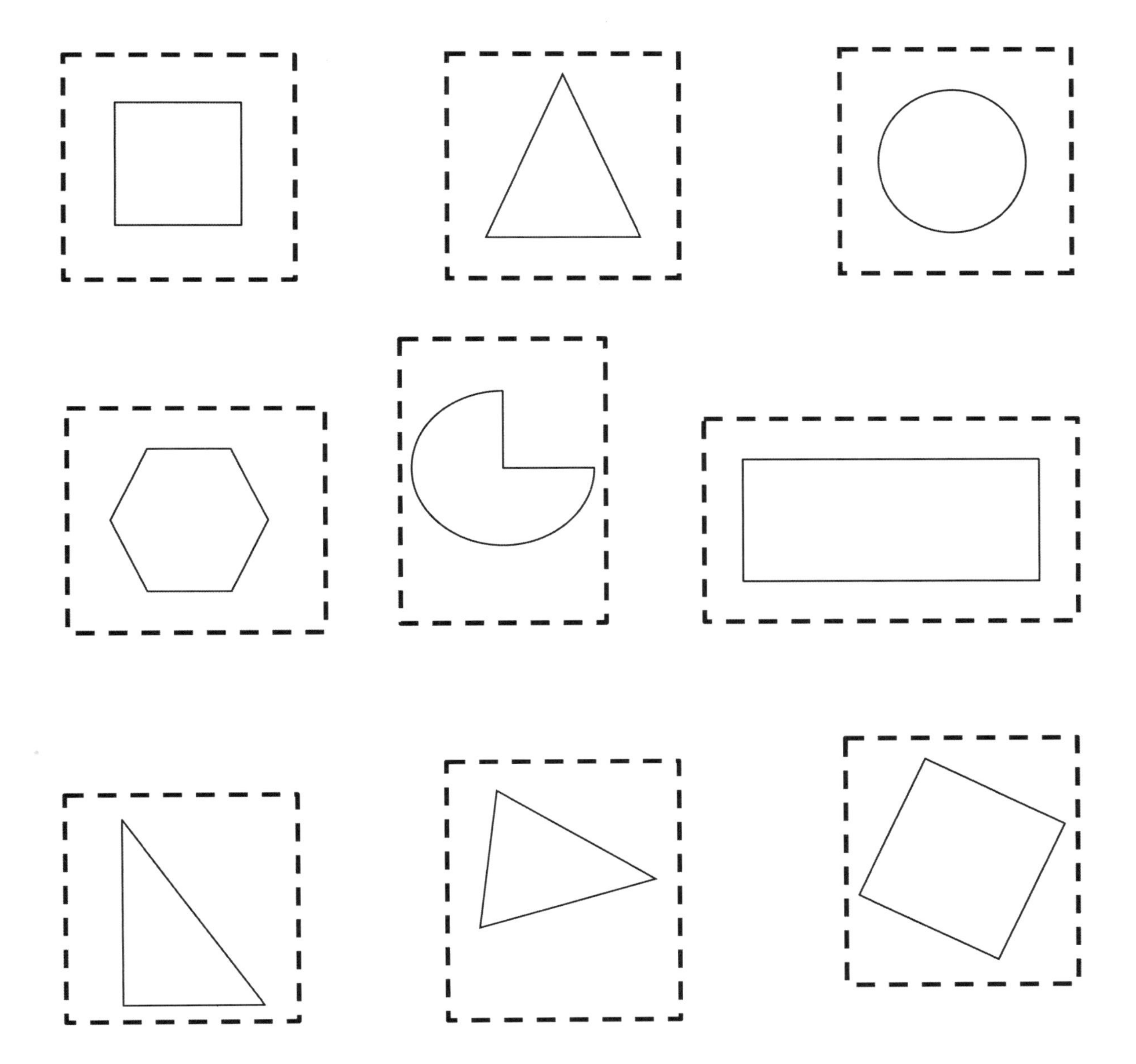

Name ______________________ Date ____________

Name ______________________ Date __________

- **Behind** the elephant, draw a shape with 4 straight sides that are exactly the same length. Color it blue.
- **Above** the elephant, draw a shape with no corners. Color it yellow.
- **In front of** the elephant, draw a shape with 3 straight sides. Color it green.
- **Below** the elephant, draw a shape with 4 sides, 2 long and 2 short. Color it red.
- **Below** the elephant, draw a shape with 6 corners. Color it orange.

On the back of your paper, draw 1 hexagon and 4 triangles.
How many shapes did you draw? Put your answer in the circle.

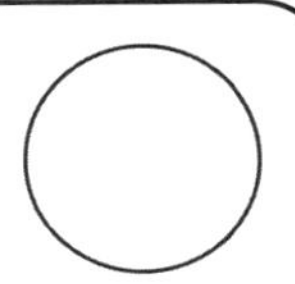

triangle

rectangle

signs

square

hexagon

signs

circle

signs

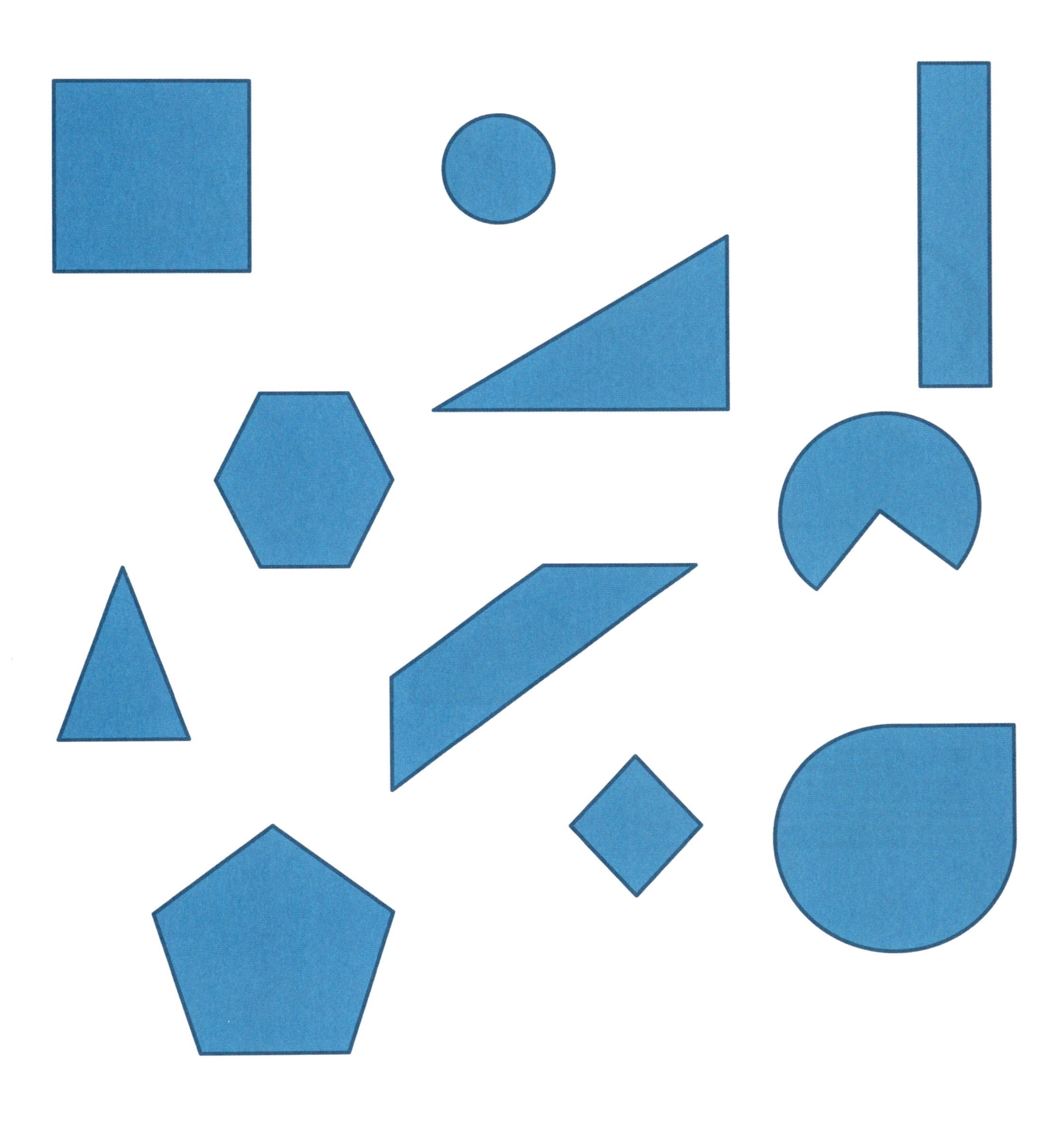

shapes

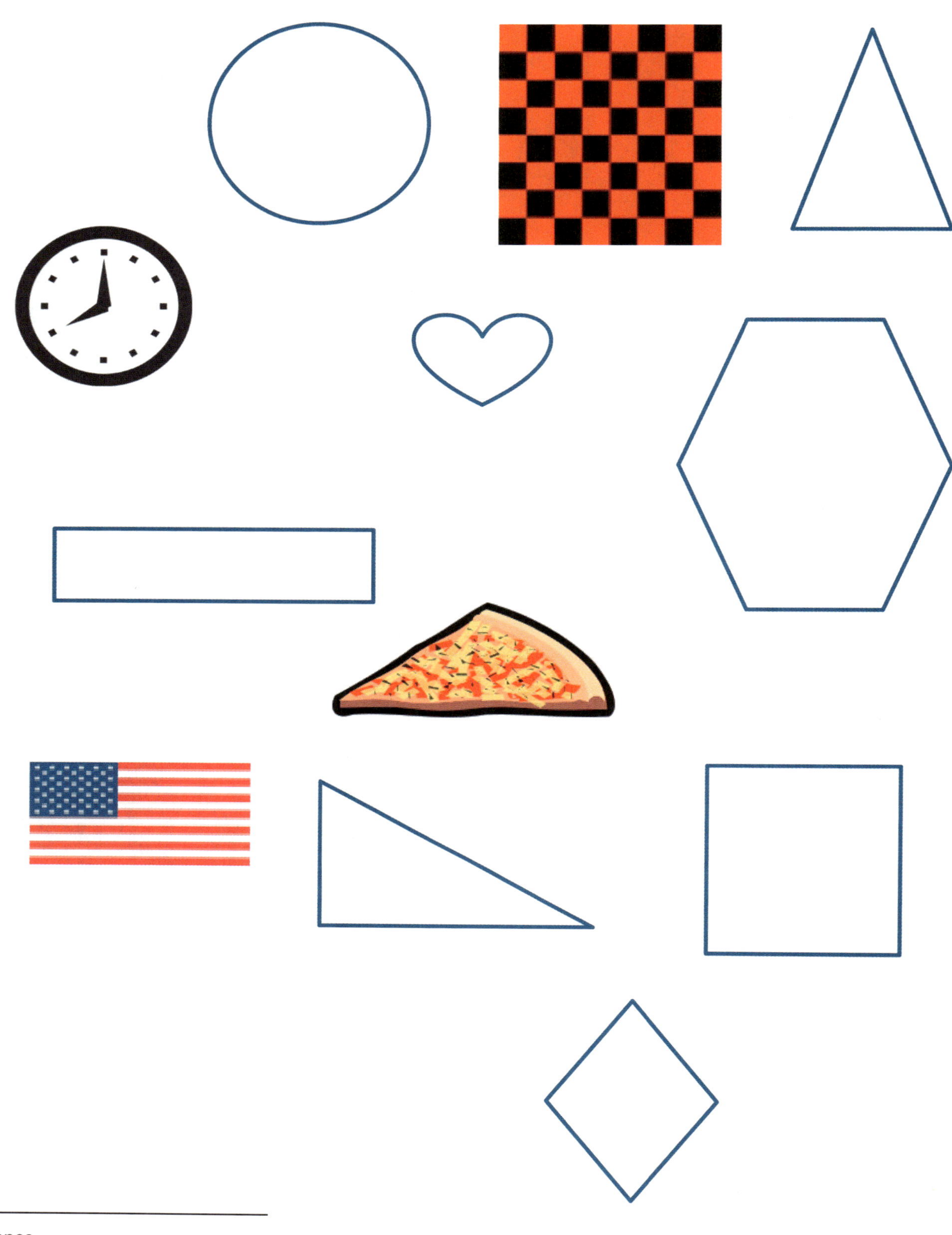

shapes

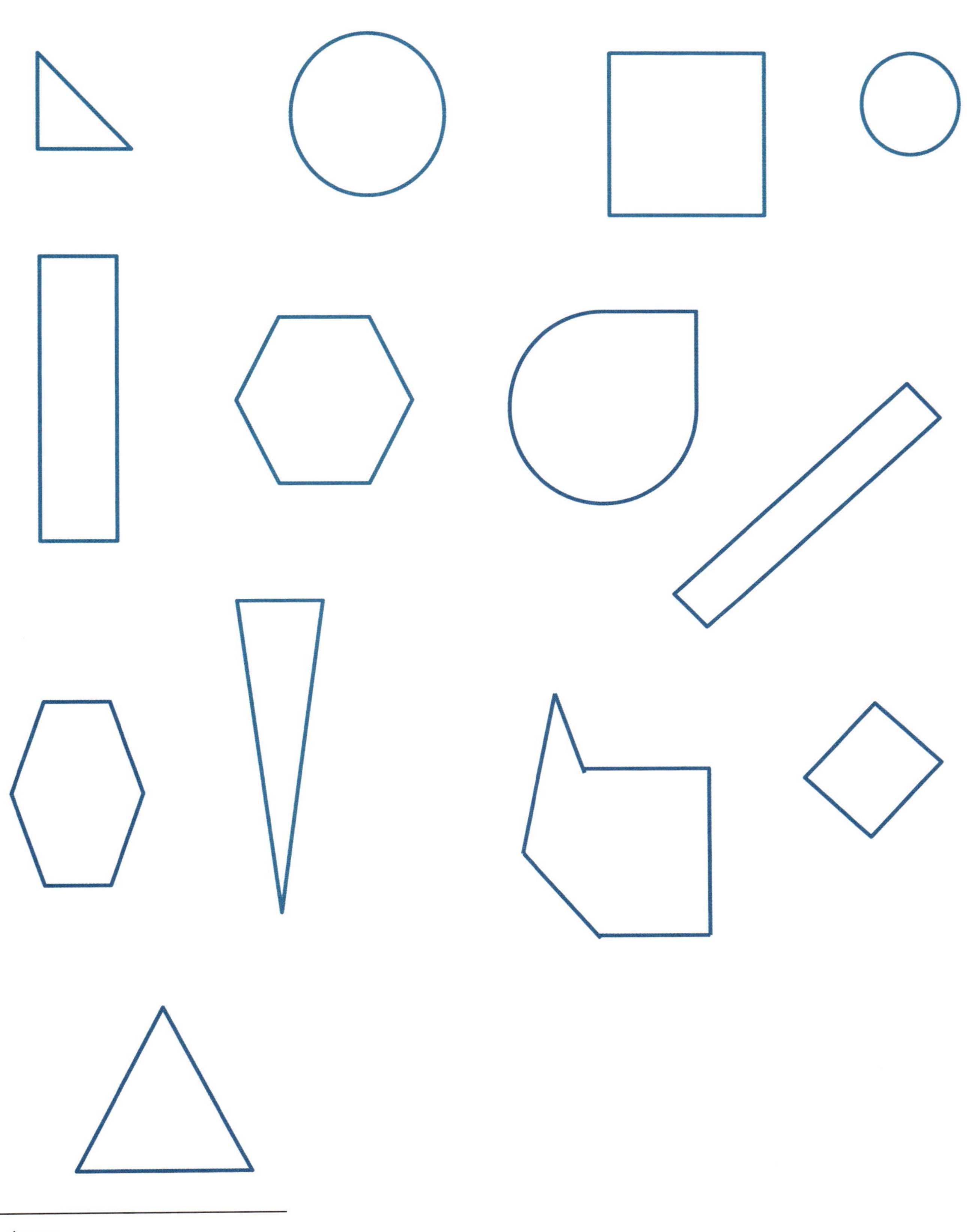

shapes

Topic B

Three-Dimensional Solid Shapes

K.G.1, K.G.2, K.G.4, K.MD.3

Focus Standards:		K.G.1	Describe objects in the environment using names of shapes, and describe the relative positions of these objects using terms such as *above, below, beside, in front of, behind,* and *next to*.
		K.G.2	Correctly name shapes regardless of their orientations or overall size.
		K.G.4	Analyze and compare two- and three-dimensional shapes, in different sizes and orientations, using informal language to describe their similarities, differences, parts (e.g., number of sides and vertices/"corners") and other attributes (e.g., having sides of equal length).
Instructional Days:		3	
Coherence	**-Links from:**	GPK–M2	Shapes
	-Links to:	G1–M5	Identifying, Composing, and Partitioning Shapes

The lessons of Topic B replicate concepts taught in Topic A but with solid shapes. Lesson 6 begins with students finding solid shapes in their environment. They might find bottles of paint, tissue boxes, balls, or crayons and describe these objects to their neighbor using informal language. "My ball is round, and it bounces!" "This tissue box has a lot of pointy corners." Some students might even use the flat shape vocabulary they learned in Topic A to describe their solid shape. "There are a lot of rectangles on my tissue box, too."

In Lesson 7, students learn the names of the solid shapes and focus on their attributes. They are asked to explain their thinking as they classify the solid shapes into categories. "I'm putting the cube and rectangular prism together because they have six sides." "The sphere and cylinder roll. They should go together." Lesson 8 guides the students to use their new solid shape lexicon to communicate the position of solid shapes to each other. Students identify, name, and position shapes relative to each other.

GK-M2-TE-B2-1.3.1-01.2016

A Teaching Sequence Toward Mastery of Three-Dimensional Solid Shapes	
Objective 1:	**Find and describe solid shapes using informal language without naming. (Lesson 6)**
Objective 2:	**Explain decisions about classification of solid shapes into categories. Name the solid shapes. (Lesson 7)**
Objective 3:	**Describe and communicate positions of all solid shapes using the words *above, below, beside, in front of, next to*, and *behind*. (Lesson 8)**

Lesson 6

Objective: Find and describe solid shapes using informal language without naming.

Suggested Lesson Structure

Fluency Practice	(12 minutes)
Application Problem	(5 minutes)
Concept Development	(25 minutes)
Student Debrief	(8 minutes)
Total Time	**(50 minutes)**

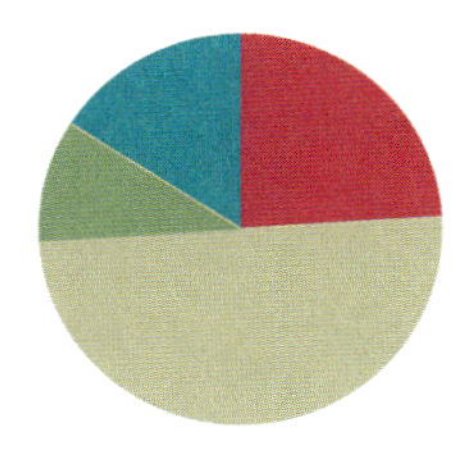

Fluency Practice (12 minutes)

- Beep Number **K.CC.4a** (4 minutes)
- Hide and See 5 **K.OA.2** (4 minutes)
- Take Apart Groups of Circles **K.OA.1** (4 minutes)

Beep Number (4 minutes)

Note: This fluency activity ensures that students gain flexibility with number order in both directions on the number line.

Materials: (T) Personal white board (optional) (S) Number path (Fluency Template) (optional)

T: Let's play Beep Number! Listen carefully while I count. Instead of saying a number, I'll say *beep*. You can touch each number on your number path as I say it. When you know what the beep number is, raise your hand. 5, beep, 7. (Wait until all hands are raised, and then give the signal.)

S: 6.

T: (Turn over the personal board to reveal the number 6 so that students can verify that their answer was correct.)

T: 7, beep, 5. (Wait until all hands are raised, and then give the signal.)

S: 6.

T: (Turn over the personal board to reveal the number 6.)

Continue in a thoughtful sequence, intermingling counting up and counting down. Return to a simpler sequence if students have difficulty.

The teacher's use of the personal white board is optional, but it can increase engagement as students perceive the number as secret. Initially, students may rely heavily on the number path in order to determine the missing number. Challenge students to solve mentally when they are ready.

Hide and See 5 (4 minutes)

Materials: (S) 5 linking cubes, personal white board

Note: In this activity, students' understanding of the conservation of a number develops into part to whole thinking at the concrete level, anticipating the work of Module 4 (number bonds, addition, and subtraction).

T: Touch and count your cubes.
S: 1, 2, 3, 4, 5.
T: Hide 2 behind your back. How many can you see?
S: 3.
T: Put them back together. How many cubes do you have?
S: 5.
T: Hide 1 behind your back. How many can you see?
S: 4.
T: Put them back together. How many cubes do you have?
S: 5.

Variation: As students put the cubes together, they can write the expressions on their personal white boards. Challenge students to list all possible combinations.

Take Apart Groups of Circles (4 minutes)

Materials: (S) Personal white board

Note: In order to meet the goal of adding and subtracting fluently within 5, students need to begin practicing early and regularly.

T: Draw three circles on your board. (Wait for students to do this.) Put an X on two of them. How many circles have an X?
S: 2.
T: How many circles do not have an X?
S: 1.
T: How many circles are on your board?
S: 3.
T: Raise your hand when you can say the number sentence starting with 2. (Wait for all students to raise their hands, and then signal.) Ready?
S: 2 and 1 make 3.
T: Very good. Let's go a little faster now. Erase. Draw four circles on your board. (Wait for students to do this.) Put an X on three of them. (Wait.) How many do not have an X?
S: 1.
T: Raise your hand when you can say the number sentence starting with 3. (Wait for all students to raise their hands, and then signal.) Ready?
S: 3 and 1 make 4.

Continue working through problems within 5. Alternatively, students can write the equation when 3 is the total and the expressions when 4 or 5 is the total.

Application Problem (5 minutes)

Have students work with a partner. Give each set of students a small ball and a cube.

We are going to do a test. Take turns with your partner. Roll the ball back and forth between you a few times. Watch the ball carefully as it rolls. Now, try to roll the block between you. Talk to your partner about what happens. Why do you think the objects behave so differently? What would be the best way to get the block to your partner? Why don't cups that have a circle on the bottom roll off the table?

Note: This Application Problem requires students to start thinking about the differences between balls and cubes in preparation for today's lesson.

Concept Development (25 minutes)

Preparation: As with the hexagons, prior to the lesson, strategically place some extra examples of the geometric solids around the classroom if they are not already present. Suggestions include party hats, cans, snow cone cups, drums, and boxes.

Materials: (S) 1 bag containing a set of geometric solids per student pair (solids should include a cone, a cylinder, a cube, and a sphere), clipboard, paper, pencil, real or toy magnifying glass (if available)

T: I have something new for you to explore today! You will be working with your partner. Please take everything out of your bag. I will give you a few minutes to look and talk with your partner about what you notice.

S: (Allow three minutes for free exploration and discussion time.)

T: Place your things on your desk. Stand up and look down at them as though you were a bird. What do you notice?

S: From up here, this one looks like a square! → This one looks like a circle. → From here, these two look alike.

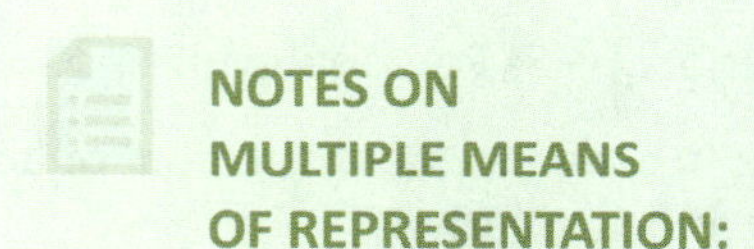

To help English language learners, introduce key vocabulary like *flat, solid, edges, corners, sides, points, curves,* and *solids* days before you teach the lesson. Post these terms on the word wall with visuals or a concrete example of each.

T: Now, pretend that you are an ant. Bend down and look from eye level across the top of your desk. When we did this with your flat shapes, you said you couldn't see them anymore. What happens this time?

S: They stick up! → Now, I see a triangle. → They are not flat.

T: You're right. They are not flat. We call these **solids**. Find the solid that looks like this. (Hold up the sphere.) Tell me about this solid.

S: It looks like a ball. → It is round.

MP.6

T: (Hold up the cube.) Look at this solid. Find the one that looks like it on your desk. How is it different? (Continue examining the solids until students have had a chance to describe them all. Encourage the students to use language such as *edges, corners, sides, points*, and *curves* in their discussion.)

T: Put your shapes back in the bag. Take out your detective materials. You and your partner are going to hunt for these shapes around our classroom. When you find one, draw it on your paper. (Allow students five minutes to identify some of the solids in the environment.)

T: Please return to your seats. Would anyone like to show and share about what they found? (Allow time for discussion and sharing.) We will find some more solids in our Problem Set.

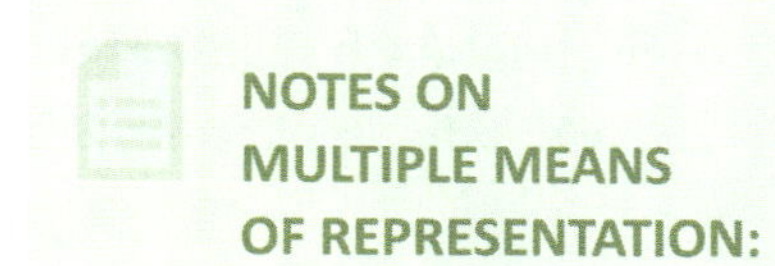

NOTES ON MULTIPLE MEANS OF REPRESENTATION:

Students working below grade level may experience difficulty with the sorting activity. To assist them, use interactive technology where students are challenged to sort solids and two-dimensional shapes.

Problem Set (10 minutes)

Students should do their personal best to complete the Problem Set within the allotted time.

Note: If students finish early, have them draw solid shapes that they see around them.

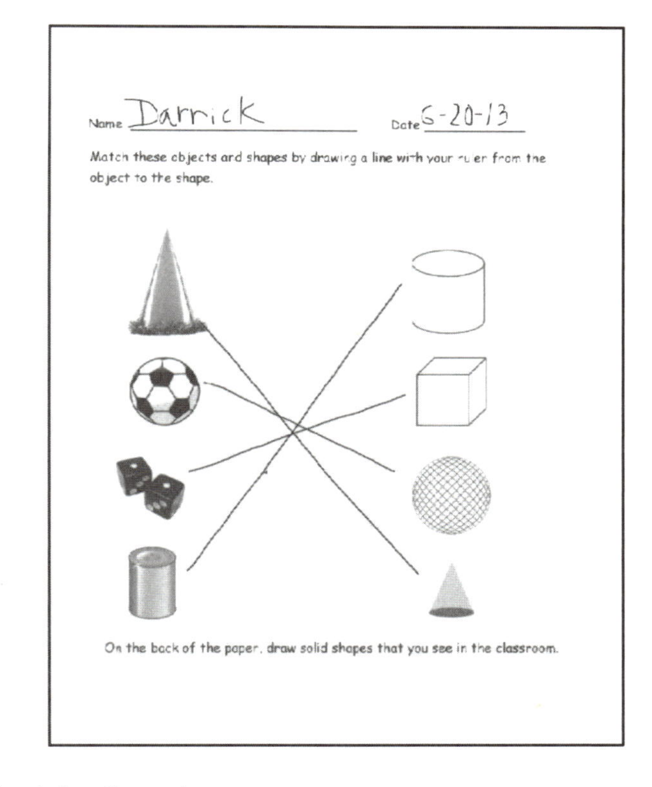

Name Darrick Date 5-20-13

Match these objects and shapes by drawing a line with your ruler from the object to the shape.

On the back of the paper, draw solid shapes that you see in the classroom.

Student Debrief (8 minutes)

Lesson Objective: Find and describe solid shapes using informal language without naming.

The Student Debrief is intended to invite reflection and active processing of the total lesson experience.

Invite students to review their solutions for the Problem Set. They should check work by comparing answers with a partner before going over answers as a class. Look for misconceptions or misunderstandings that can be addressed in the Debrief. Guide students in a conversation to debrief the Problem Set and process the lesson.

Any combination of the questions below may be used to lead the discussion.

- What **solids** did you match that were curved? What solids did you match that were not curved?
- Which shapes were the hardest to match? Why?
- Explain to your partner what you drew on the back of your paper. Can you think of other objects around you that are these solid shapes? Have a volunteer or two share their drawings.

- What new (or significant) math vocabulary did we use today to communicate precisely? How can you tell about each solid without using the solid's name?
- How did the Application Problem connect to today's lesson?

Name ______________________________ Date ______________

Match these objects and solids by drawing a line with your ruler from the object to the solid.

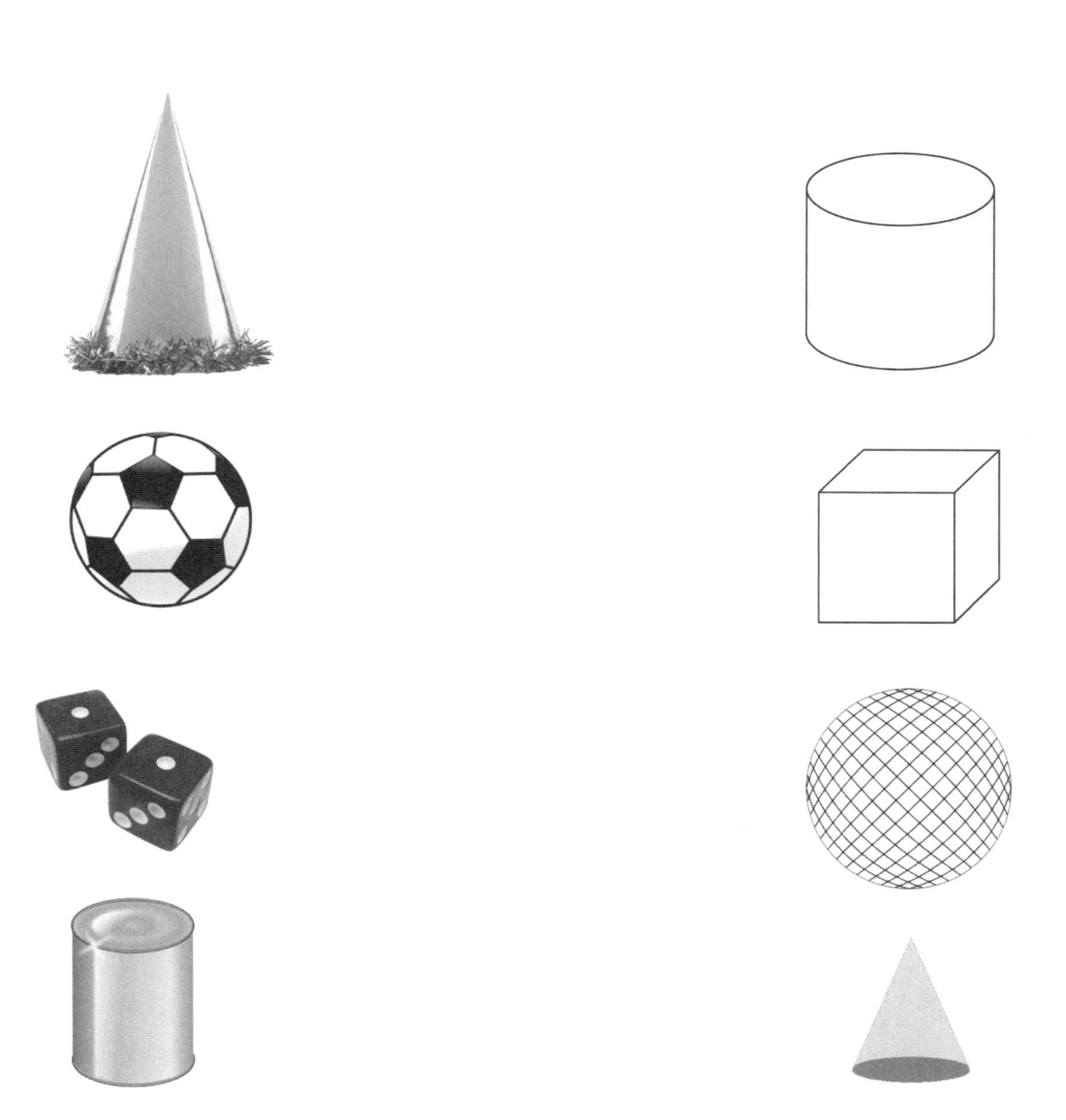

On the back of the paper, draw solid shapes that you see in the classroom.

Name ________________________________ Date ______________

Find things in your house or in a magazine that look like these solids. Draw the solids or cut out and paste pictures from a magazine.

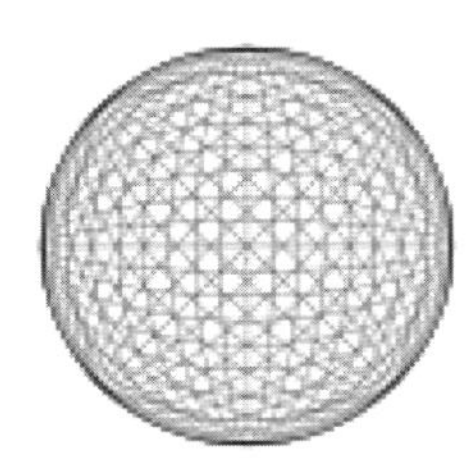

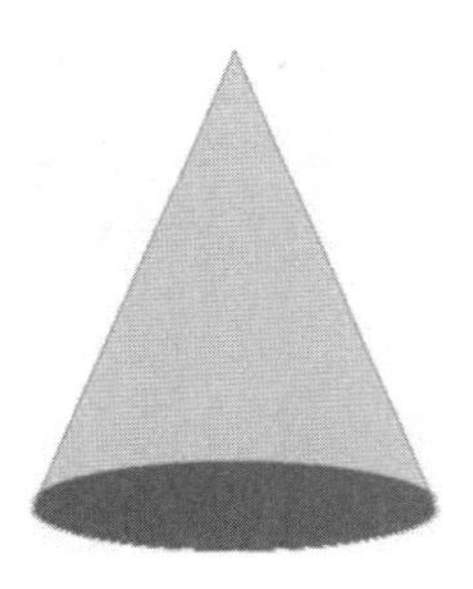

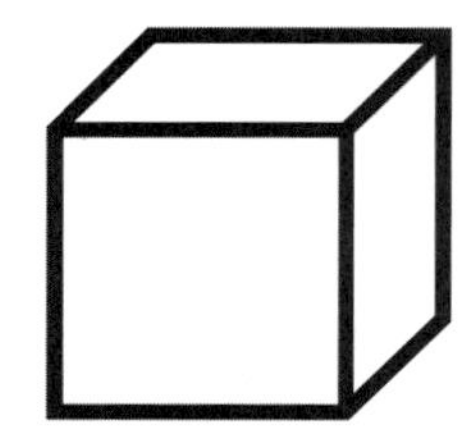

1	2	3	4	5	6	7	8	9	10

1	2	3	4	5	6	7	8	9	10

1	2	3	4	5	6	7	8	9	10

1	2	3	4	5	6	7	8	9	10

number path

Lesson 7

Objective: Explain decisions about classification of solid shapes into categories. Name the solid shapes.

Suggested Lesson Structure

- Fluency Practice (12 minutes)
- Application Problem (5 minutes)
- Concept Development (25 minutes)
- Student Debrief (8 minutes)

Total Time (50 minutes)

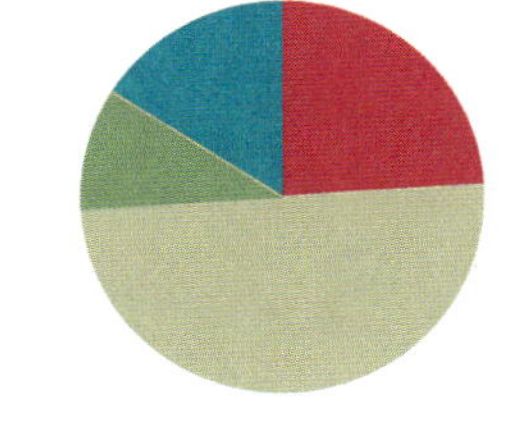

Fluency Practice (12 minutes)

- Show Me Shapes **K.G.2** (4 minutes)
- Making 5 with 5-Group Mats **K.OA.1** (5 minutes)
- 5-Group Hands **K.CC.2** (3 minutes)

Show Me Shapes (4 minutes)

Materials: (S) Assortment of solid shapes, possibly a mixture of everyday objects and wooden or plastic solid shapes

Note: In this activity, students continue to analyze solid shapes to gain fluency with recognizing attributes and using geometric vocabulary.

Scatter the solid shapes and objects onto the students' tables or in the center of the rug.

T: Look at the shapes that are on the rug. I will ask you to find a certain kind of shape. When you find it, hold it up. Ready? Show me shapes that have points.

S: (Hold up cubes and cones.)

T: Yes. Put them back on the rug, and listen to what I want you to find next. Show me shapes that have no points.

S: (Hold up spheres.)

T: Yes. Now, show me shapes that have a curve.

S: (Hold up spheres, cones, and cylinders.)

Continue having students test each other so they practice the vocabulary.

Making 5 with 5-Group Mats (5 minutes)

Materials: (S) 5-group mats (Lesson 1 Fluency Template 1), 5 linking cubes

Note: In this activity, students work toward fluency with numbers within 5.

Conduct the activity as outlined in Lesson 1, but now have students rotate their mats so that they work with 5-groups in the vertical orientation.

5-Group Hands (3 minutes)

Materials: (T) Large 5-group cards (Lesson 1 Fluency Template 3)

Note: Students' facility with their hands lays the foundation for the use of the number line.

Conduct the activity as outlined in Lesson 1, but now continue to 10. Consider showing the cards in the vertical orientation so that students can gain flexibility in locating the 5-group.

Application Problem (5 minutes)

Materials: (S) Small piece of modeling clay

Think about the solids you investigated yesterday. Now, listen to the riddle, and make this mystery solid with your clay: I am a solid that can roll. I don't have any corners. I have zero edges. Make me!

When you are done, show the solid to your friend. Do your solids look alike?

Note: The purpose of this problem is to remind students of the vocabulary used in yesterday's lesson. They need to focus on descriptions of this type in today's lesson. It also gives kinesthetic learners a chance to manipulate the clay.

NOTES ON MULTIPLE MEANS OF ACTION AND EXPRESSION:

As the vocabulary terms *cone, face, cube, corners*, and *edges* come up in the lesson, the teacher can use gestures like touching her face and then the *face* of the solid while saying the word *face* in order to enrich English language learners' experience and make it easier for them to access the content of the lesson.

Concept Development (25 minutes)

Materials: (S) Set of geometric solids including a cube, sphere, cone, and cylinder per student pair; paper and colored pencils; small smiley face stickers

Note: In the context of polyhedra, faces must be polygonal. However, in more general contexts, a face may be circular (such as the base of a right circular cylinder) or even irregular. It is this more inclusive interpretation of face that is used in this Kindergarten module.

T: Take your solids out of your bag. We are going to look at them carefully to see if any of them have things in common. If they do, we can sort them. Does anyone have any ideas?

S: This one rolls, but these two don't. → These both have flat sides. → These have pointy parts.

T: I hear some good ideas! We will try some of them. (Hold up a **cone**.) This solid is called a cone. What do you notice about this solid?

S: It is flat on the bottom. → There is a circle on the bottom.

T: The circle, the flat part of the cone, is called a **face**. Take a smiley sticker and put it on the face of the cone. Do you have other solids that have faces?

S: This one! (Hold up a **cube**.)

T: Yes, that solid has many faces! It is called a cube. Put a smiley sticker on each face of the cube. How many faces does it have? (Continue to hold up the solids and mark the faces, counting the faces of each. Introduce the students to the names of each of the solids.)

MP.7

T: Can we sort our solids into groups of those with a face and those without?

S: Yes! (Sort the solids. Name the **sphere** and **cylinder**.)

Guide children to sort solids several times by other criteria, for example, those that roll and those that only slide, those that can stack and those that cannot, those that have corners, those that have edges, those that look like circles from above, and so on. While monitoring students as they sort, use and encourage correct vocabulary to reinforce learning.

T: Which of your solids has the most faces?

S: The cube.

T: Put your cube on one of its faces onto your piece of paper. Use your favorite colored pencil to trace around the solid. Now, lift your solid. What do you see underneath?

S: A square.

T: The face of the cube is a flat square. I wonder what would happen if you traced the face of your cone? (Allow students to try.)

S: It makes a circle.

T: Should we trace a face of the cylinder?

S: Yes! It makes a circle, too.

T: Trace the faces of any of your objects to make shape designs on your paper. (Allow time for tracing and discussion.)

T: Put your solids away. Would anyone like to share their Trace the Face picture, and tell us how the shapes are the same and how they are different?

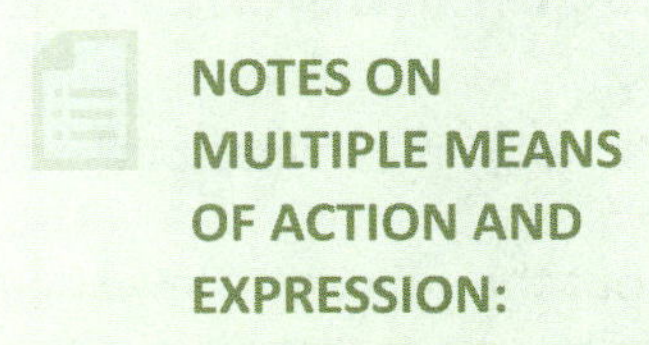

NOTES ON MULTIPLE MEANS OF ACTION AND EXPRESSION:

Facilitate struggling students' talk by providing them with various sentence frames such as, "These two solids are the same because they have..." and, "This one rolls, but this one just...." Encourage them to use the names of the solids.

Problem Set (10 minutes)

Students should do their personal best to complete the Problem Set within the allotted time.

Student Debrief (8 minutes)

Lesson Objective: Explain decisions about classification of solid shapes into categories. Name the solid shapes.

Invite students to review their solutions for the Problem Set. They should check work by comparing answers with a partner before going over answers as a class. Look for misconceptions or misunderstandings that can be addressed in the Debrief. Guide students in a conversation to debrief the Problem Set and process the lesson.

Any combination of the questions below may be used to lead the discussion.

- Which objects did you circle that were **cylinders**? (**Cubes, cones**, and **spheres**.)
- What did you need to remember when you were finding the cylinders to circle? (Cubes, cones, and spheres.) Did anyone think of something else?
- What new (or significant) math vocabulary did we use today to communicate precisely? (Emphasize **faces**, corners, and edges.)

- How can you tell about each shape without using the shape's name?
- How did the Application Problem connect to today's lesson?
- What were some different ways we sorted our shapes?

Name ______________________________ Date ______________

Circle the cylinders with red.

Circle the cubes with yellow.

Circle the cones with green.

Circle the spheres with blue.

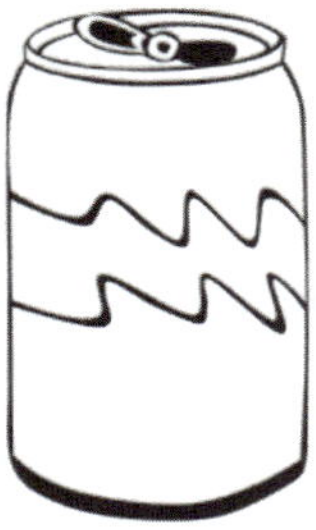

Name ______________________________ Date ______________

Cut one set of solid shapes. Sort the 4 solid shapes. Paste onto the chart.

These have corners.	These do not have corners.

Cut the other set of solid shapes, and make a rule for how you sorted them. Paste onto the chart.

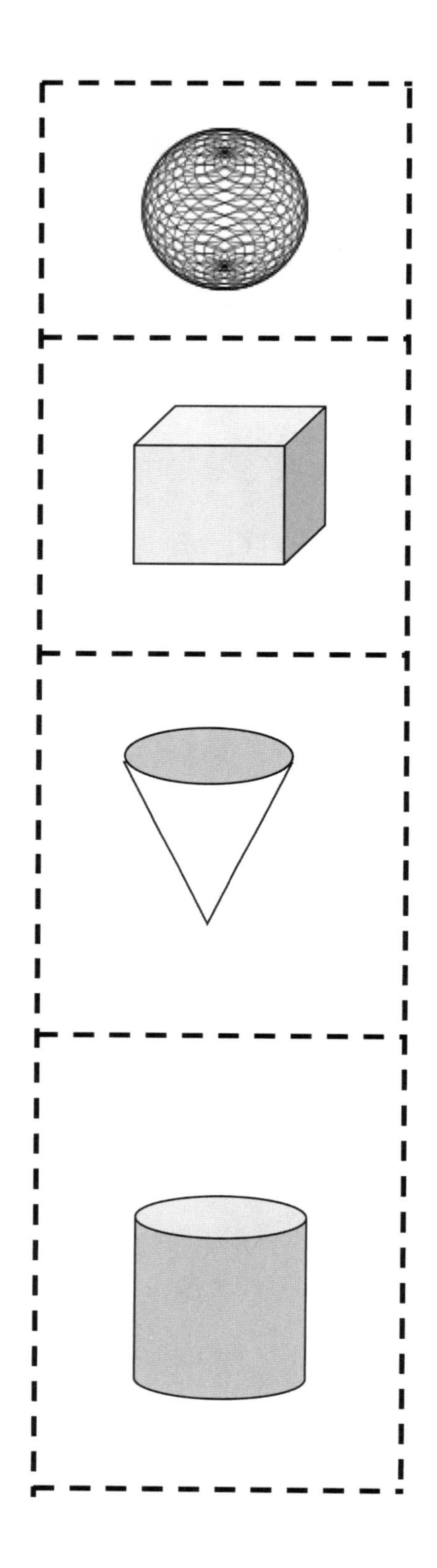

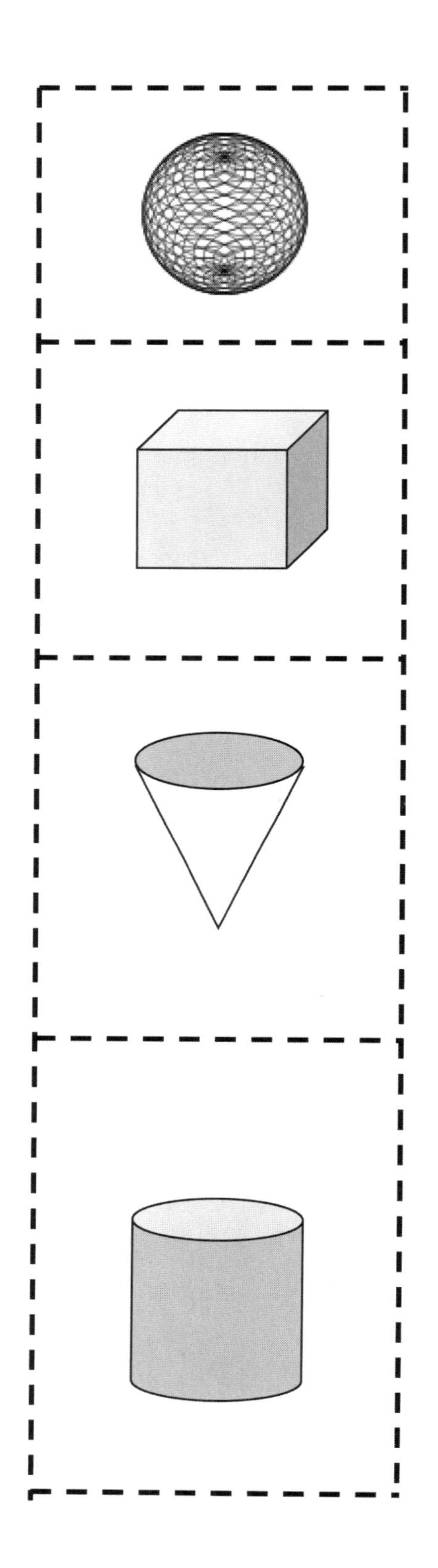

Lesson 8

Objective: Describe and communicate positions of all solid shapes using the words *above, below, beside, in front of, next to*, and *behind*.

Suggested Lesson Structure

■ Fluency Practice	(12 minutes)
■ Application Problem	(5 minutes)
■ Concept Development	(26 minutes)
■ Student Debrief	(7 minutes)
Total Time	**(50 minutes)**

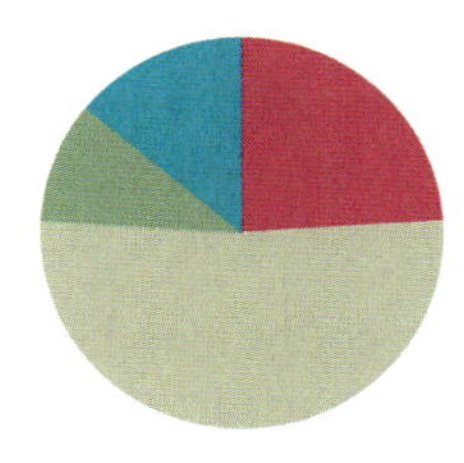

Fluency Practice (12 minutes)

- Position Words Game **K.G.1** (4 minutes)
- Show Me Shapes **K.G.2** (4 minutes)
- Rekenrek Roller Coaster **K.CC.4a** (4 minutes)

Position Words Game (4 minutes)

Note: As students work with position words, they are analyzing their world and their relationship to that world in space. This leads to insights about measurement and distance.

Call students to the rug with their partners. Use position words such as *above, below, beside, in front of, next to*, and *behind* to give directions for activities to do with a partner. For example, "Partner A, put your hand above Partner B's head. Stand beside your partner. Partner B, stand in front of Partner A."

The benefit of partner work is that they check and correct each other and, in the process, explain their thinking.

Show Me Shapes (4 minutes)

Note: Now that students have had the chance to really study the attributes of shapes, they should visualize each shape as they say the name of it.

Conduct the activity as outlined in Lesson 7, but now use the names of solid shapes.

Rekenrek Roller Coaster (4 minutes)

Materials: (T) 20 Rekenrek

Note: As students gain deeper understanding of the numbers in relationship to 5, the Rekenrek allows them to start building a relationship to 10 ones, as outlined in Kindergarten Module 1 Lesson 23.

T: Let's practice counting with the Rekenrek. (Show students the 20 Rekenrek with the side panel attached.) Say how many you see. (Slide the balls you want the students to count completely to one side.)

Direct the students to gradually raise their hands as the numbers increase and to lower their hands as the numbers decrease, mimicking the motion of a roller coaster. A suggested sequence is counting up, counting down, and then in short sequences, 1, 2, 3, 2, 3, 4, 3, 4, 5, 4, 3, etc. Gradually build up to 10.

Be careful not to mouth the number words or count along with the students. Listen carefully for hesitations or errors, and return to a simpler sequence if necessary. If students demonstrate mastery, consider introducing the 5-group orientation (e.g., 6 as 5 red beads on top and 1 red bead on the bottom).

Application Problem (5 minutes)

Materials: (S) Small ball of clay

Make a sphere with your ball of clay. Make your ball into a cylinder. Make your cylinder into a cube. Make your cube into a cone. Put your cone next to your partner's. Partner A, put your cone above Partner B's.

Note: This Application Problem reviews vocabulary from yesterday's lesson and bridges to the work of positioning solids in today's lesson.

Concept Development (26 minutes)

Materials: (T) Set of geometric solids in a paper bag; set of flash cards in a paper bag showing the words *above, beside, below, in front of, next to*, and *behind* (S) Set of geometric solids per pair

T: We are going to play a math game today called Guess What I Am. I'm going to call on a lot of volunteers, so be ready to be mathematicians! I have two bags in front of me. Who can guess what is in my bags? (Shake bags and elicit guesses from students.)

T: Student A, please come up to help me. I want you to put your hand in this bag and find one of the objects, but don't look at it! See if you can guess what it is just by feeling it. Here is a hint: It is something that we looked at yesterday. (Allow student to locate one of the solids in the bag.) Tell us about it.

S: It feels smooth and round. It is a sphere!

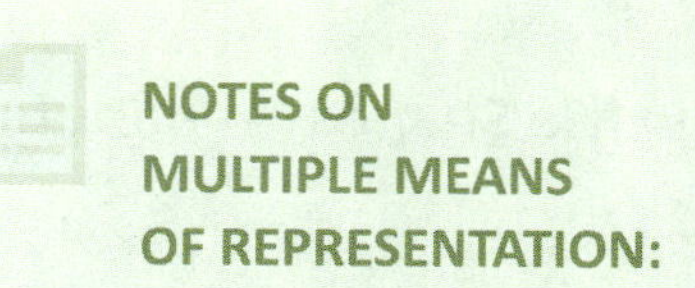

NOTES ON MULTIPLE MEANS OF REPRESENTATION:

English language learners are able to participate better if the flash cards have a visual illustrating what *beside, above, in front of, next to*, and *behind* mean. This type of multiple representation helps them learn these terms more quickly.

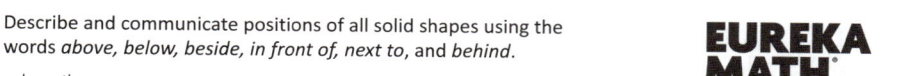

T: Take it out of the bag. Is he right? (Yes.) Find your spheres, and put them on your desks. (Allow time for pairs of students to put their spheres on their desks.)

T: Student B, would you help me next? Find something in the bag, and see if you can tell us what it is without looking. What do you feel?

S: I feel something with lots of corners. It has lots of flat sides. It is a cube!

MP.3

T: Is he right? (Yes!) Now, Student B, take a card out of the other bag. (Student selects a card.) This card says *beside*. Find a cube in your bag, and put it beside your sphere. Student C, it's your turn!

S: I feel something that is flat on one side and pointy on the other. It has one face. It is a cone. (Student takes it out of the bag.) I'm right!

T: Now, choose a card. (Student selects a card.) The card says *above*. Find your cones, and put them above the cube. Look! You made a building!

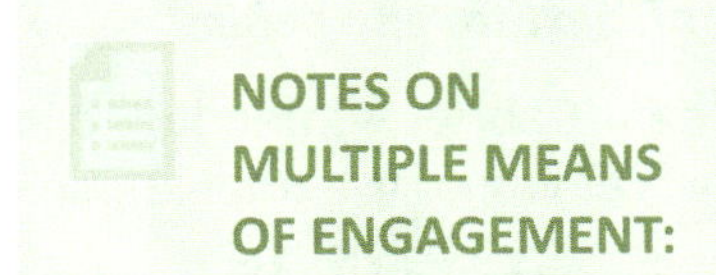

NOTES ON MULTIPLE MEANS OF ENGAGEMENT:

Challenge students working above grade level by handing them interesting images and getting them to analyze them from shape and position with a partner. Have them present their insights in the Debrief.

Play the game at a fast pace for several minutes or until all students have had a turn to identify a solid by touch. Put solids or cards back in the bags as often as necessary to continue the exercise. If the combinations create impossible situations, such as *put the sphere above the cone*, ask students what positional word could be used instead.

T: Now, arrange the solids on your desk. You will play a similar game with your partner, but in a different way. Tell your partner, "I am the solid that is next to the cube. What am I?" When your partner guesses the solid correctly, it will be his turn to give you a riddle. (Allow students time to practice using the words of position and identifying the solids.)

T: Put your solids away in your bag. Put your bag in front of you. I will put your Problems Sets beside them.

Problem Set (10 minutes)

Students should do their personal best to complete the Problem Set within the allotted time.

Note: To extend this activity, the teacher could copy a second set of shapes and use these clues.

- Paste the shape with no faces above the train.
- Paste the shape with many faces behind the train.
- Paste the shape with two faces in front of the train.
- Paste the shape with one point below the train.

Student Debrief (7 minutes)

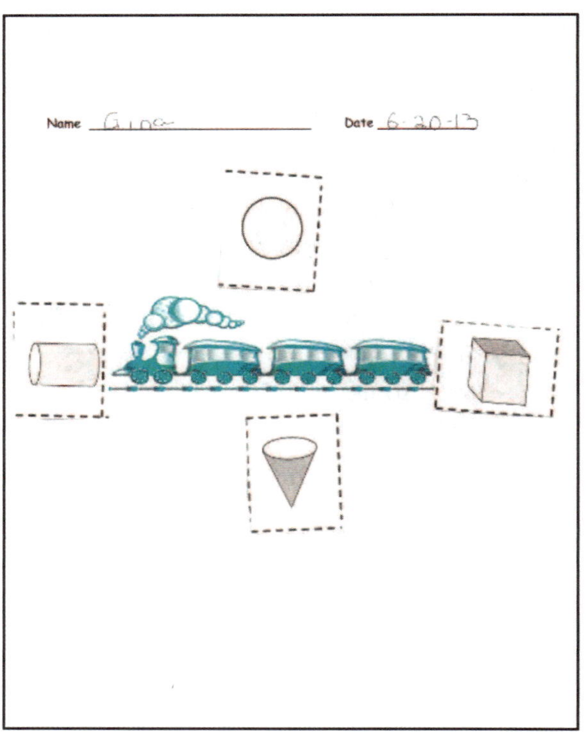

Lesson Objective: Describe and communicate positions of all solid shapes using the words *above, below, beside, in front of, next to*, and *behind*.

The Student Debrief is intended to invite reflection and active processing of the total lesson experience.

Invite students to review their solutions for the Problem Set. They should check work by comparing answers with a partner before going over answers as a class. Look for misconceptions or misunderstandings that can be addressed in the Debrief. Guide students in a conversation to debrief the Problem Set and process the lesson.

Any combination of the questions below may be used to lead the discussion.

- What new (or significant) math vocabulary did we use today to communicate precisely?
- Where did you place each solid on your paper? (Go through each direction, and compare where students put their shapes on their paper.)
- Were there important words you needed to know to complete this Problem Set?
- Compare with your partner. Did you put your shapes in the same place as your partner?
- What shapes do you see on your paper?
- How did the Application Problem connect to today's lesson?

Name ______________________________ Date ______________

Directions: Read to students.

Paste the sphere **above** the train.
Paste the cube **behind** the train.
Paste the cylinder **in front of** the train.
Paste the cone **below** the train.

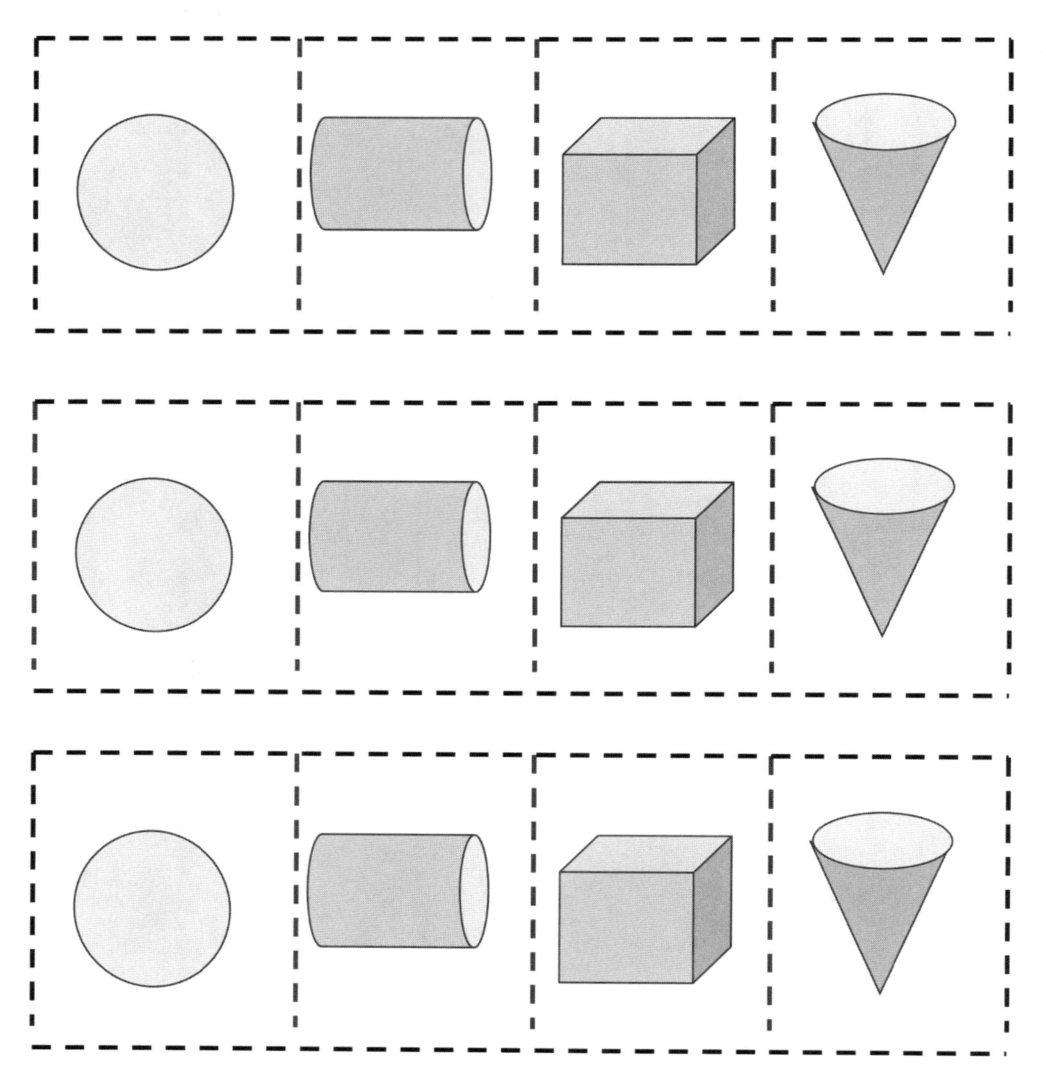

Provide one strip for every student.

Name ______________________ Date ____________

Tell someone at home the names of each solid shape.

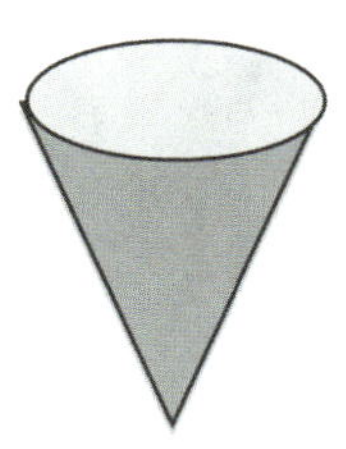

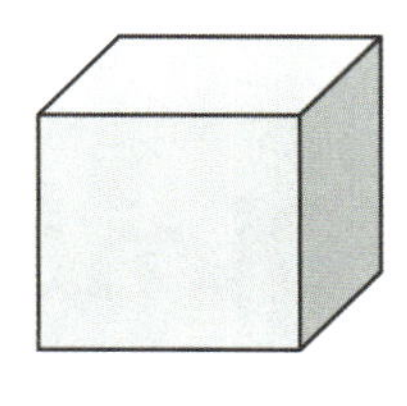

Sphere Cylinder Cone Cube

Color the car **beside** the stop sign green.

Circle the **next** car with blue.

Color the car **behind** the circled car red.

Draw a road **below** the cars.

Draw a policeman **in front of** the cars.

Draw a sun **above** the cars.

Topic C

Two-Dimensional and Three-Dimensional Shapes

K.MD.3, K.G.3, K.G.4, K.G.1, K.G.2

Focus Standards:	K.MD.3	Classify objects into given categories: count the numbers of objects in each category and sort the categories by count.
	K.G.3	Identify shapes as two-dimensional (lying in a plane, "flat") or three-dimensional ("solid").
	K.G.4	Analyze and compare two- and three-dimensional shapes, in different sizes and orientations, using informal language to describe their similarities, differences, parts (e.g., number of sides and vertices/"corners") and other attributes (e.g., having sides of equal length).
Instructional Days:	2	
Coherence -Links from:	GPK–M2	Shapes
-Links to:	G1–M5	Identifying, Composing, and Partitioning Shapes

Topic C closes the module with discrimination between flats and solids. In Lesson 9, students identify and sort flat and solid shapes. The goal of this lesson is to focus each student's attention on the attributes of a flat or solid shape instead of trusting how it looks. The students learn to sort shapes and explain the reason for their groupings.

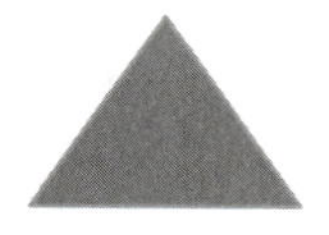

Young children might group the first and third shapes because "they look like triangles" but not the second shape because "it doesn't look like other triangles." This module closes in Lesson 10 with a culminating task that begins by asking students to distinguish between variants, non-examples, and examples of flat shapes. The task continues as students relate the flat shapes to solid shapes as they create a solid and flat shape display.

A Teaching Sequence Toward Mastery of Two-Dimensional and Three-Dimensional Shapes	
Objective 1:	**Identify and sort shapes as two-dimensional or three-dimensional, and recognize two-dimensional and three-dimensional shapes in different orientations and sizes. (Lesson 9)**
Objective 2:	**Culminating task—collaborative groups create displays of different flat shapes with examples, non-examples, and a corresponding solid shape. (Lesson 10)**

Lesson 9

Objective: Identify and sort shapes as two-dimensional or three-dimensional, and recognize two-dimensional and three-dimensional shapes in different orientations and sizes.

Suggested Lesson Structure

Fluency Practice	(10 minutes)
Application Problem	(5 minutes)
Concept Development	(25 minutes)
Student Debrief	(10 minutes)
Total Time	**(50 minutes)**

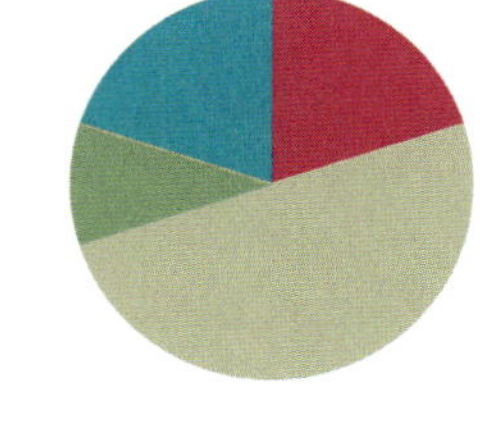

Fluency Practice (10 minutes)

- Groups of Shapes (Solid Shapes) **K.G.2** (3 minutes)
- Groups of 9 **K.CC.4b** (3 minutes)
- Hide and See 5 **K.OA.2** (4 minutes)

Groups of Shapes (Solid Shapes) (3 minutes)

Note: Kinesthetic learners benefit greatly from getting up and moving in this fluency activity. As they move, they are analyzing and are encouraged to talk about how they know where to go.

Materials: (T) Signs with pictures of shapes to indicate where to form each group (S) Assortment of real-world objects and wooden or plastic solid shapes

Conduct the activity as described in Lesson 5, but with solid shapes.

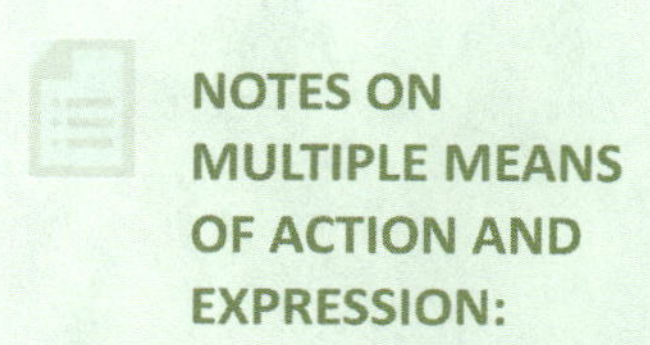

NOTES ON MULTIPLE MEANS OF ACTION AND EXPRESSION:

Challenge students working above grade level by asking them to draw two or more shapes and to construct solids that match faces of their shapes. Pair them up, and ask them to challenge each other with shapes they have not drawn.

Groups of 9 (3 minutes)

Note: This fluency activity helps students gain efficiency in counting objects in varied configurations.

Conduct the activity as outlined in Lesson 2, but with 9. Allow students to share their strategies for making groups quickly.

Hide and See 5 (4 minutes)

Materials: (S) 5 linking cubes, personal white board

Note: In this fluency activity, students' understanding of the conservation of a number develops into part to whole thinking at the concrete level, anticipating the work of Module 4 (number bonds, addition, and subtraction).

Conduct the activity as described in Lesson 6. Challenge students to list all possible combinations.

Application Problem (5 minutes)

Materials: (S) Small piece of paper, pencil, ball of clay

Draw one of the shapes that we have talked about this week. Can you make a solid with your clay that has the shape you drew as one of its faces? Share your work with your partner when you are done.

Note: In addition to serving as a review, this Application Problem requires students to think about the connections among flat shapes and solids in preparation for today's Concept Development.

Concept Development (25 minutes)

Materials: (S) Cutouts from earlier in the week, including triangles, circles, rectangles, squares, and hexagons; bag of solids including a sphere, a cylinder, a cone, and a cube (Lesson 5 Template)

T: Take out all of your flat shapes and all of your solids, and arrange them in front of you.

T: I see a lot of things on your desk! Stand up and look at your things as though you were a bird. What do you see?

S: I see flat things and solids that look like circles! → I see squares.

T: Now, bend down and look across your desk as though you were an ant. What do you notice?

S: We can only see the solid shapes. → We can't see the flat shapes now.

T: Do you think we could sort all of the things on your desk? Take a few minutes to look at all of your objects and think about what things they might have in common. (Allow time for thought and experimenting.)

T: Does anyone have a sorting rule for us to try?

S: We could put all of the things with curves over here and the things that are all straight over here.

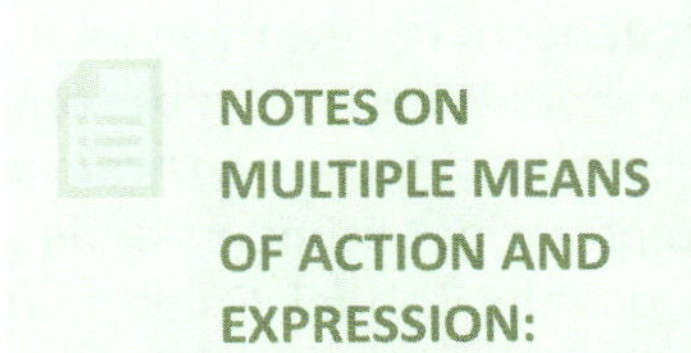

NOTES ON MULTIPLE MEANS OF ACTION AND EXPRESSION:

Students working below grade level may experience difficulty with the sorting activity. To assist them, use interactive technology to sort triangles and the more complex task of sorting multiple shapes.This kind of practice helps students improve their ability to recognize similarities and differences.

T: Good! Let's try. (Allow time for sorting; circulate to ensure accuracy.) Show your groups to your partner. Do your groups look alike? (Allow time for discussion.)

T: Did anyone think of a different rule for sorting?

S: Shapes that roll and shapes that don't. → Shapes that are flat and shapes that are solid. → Shapes with edges and shapes without edges. → Shapes with faces and shapes with no faces.

T: (Continue the sorting exercises and discussion for several minutes. Circulate to observe correct use of vocabulary and accuracy in grouping.)

T: Listen to my directions. I will say the name of a shape or a solid. When I do, echo me, find the object, and put it back in its bag. Then, I will pass out your Problem Sets.

Problem Set (10 minutes)

Students should do their personal best to complete the Problem Set within the allotted time.

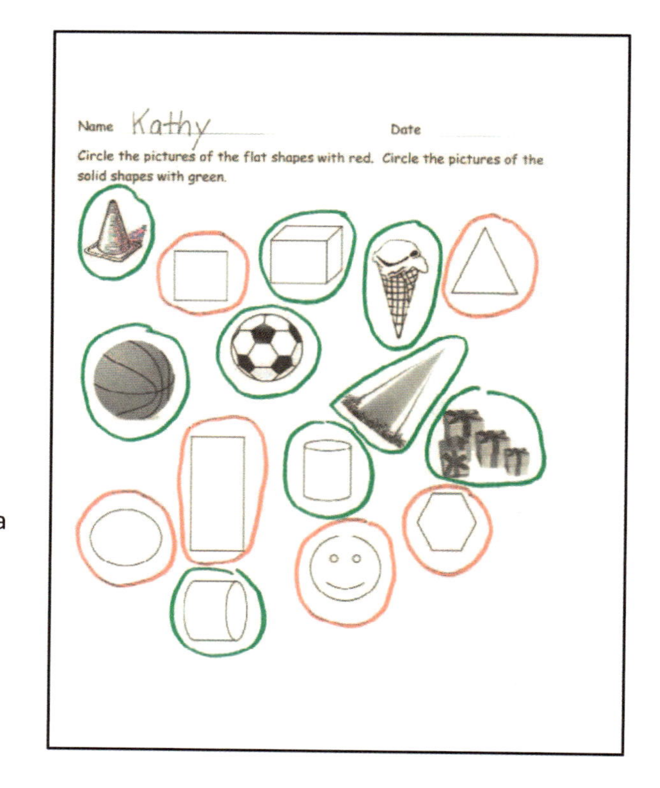

Student Debrief (10 minutes)

Lesson Objective: Identify and sort shapes as two-dimensional or three-dimensional, and recognize two-dimensional and three-dimensional shapes in different orientations and sizes.

The Student Debrief is intended to invite reflection and active processing of the total lesson experience.

Invite students to review their solutions for the Problem Set. They should check work by comparing answers with a partner before going over answers as a class. Look for misconceptions or misunderstandings that can be addressed in the Debrief. Guide students in a conversation to debrief the Problem Set and process the lesson.

Any combination of the questions below may be used to lead the discussion.

- What new (or significant) math vocabulary did we use today to communicate precisely?
- How did you determine whether to use a red or green circle? Did someone do it another way?
- Looking at your paper, who can name a flat shape? Solid shape?
- Can you name some other flat shapes that are not on your paper? Solid shapes?
- How did the Application Problem connect to today's lesson?

Name ___________________________ Date _____________

Circle the pictures of the flat shapes with red. Circle the pictures of the solid shapes with green.

Name ________________________________ Date ______________

In each row, circle the one that doesn't belong. Explain your choice to a grown-up.

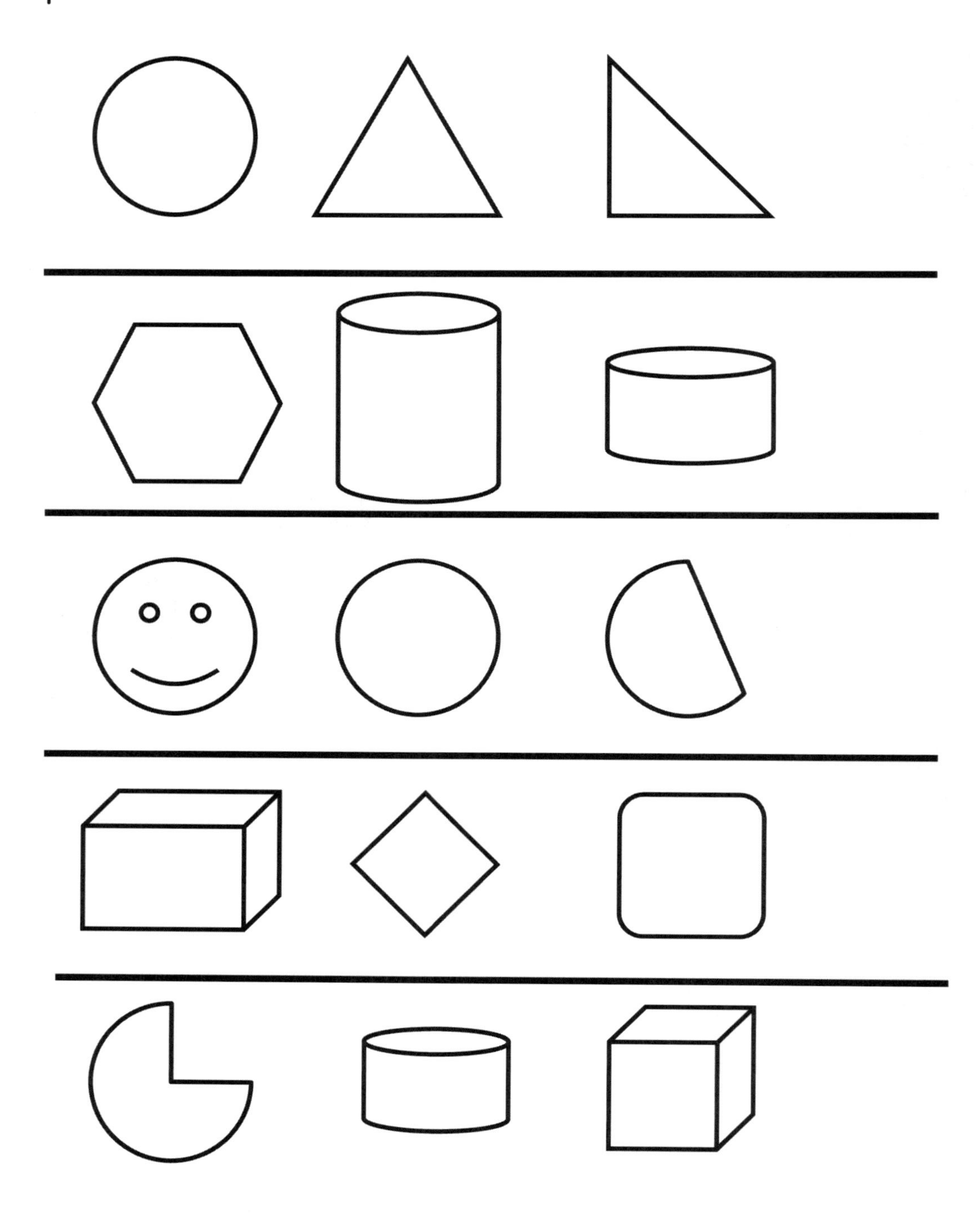

Lesson 10

Objective: Culminating task—collaborative groups create displays of different flat shapes with examples, non-examples, and a corresponding solid shape.

Suggested Lesson Structure

Fluency Practice	(5 minutes)
Concept Development	(40 minutes)
Student Debrief	(5 minutes)
Total Time	**(50 minutes)**

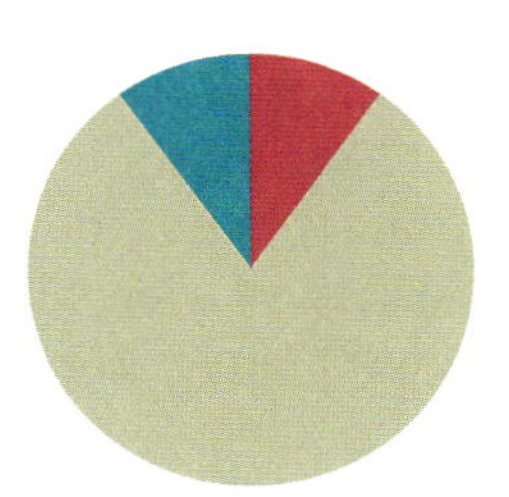

Fluency Practice (5 minutes)

- Groups of Shapes **K.G.2** (3 minutes)
- 5-Group Hands **K.CC.2** (2 minutes)

Groups of Shapes (3 minutes)

Note: The concept of a group is foundational for the concept of a unit. There are seven objects, but it is one group, one unit; conduct the activity as in Lesson 5, but with solid shapes.

5-Group Hands (2 minutes)

Materials: (T) Large 5-group cards (Lesson 1 Fluency Template 3)

Conduct the activity as outlined in Lesson 1, but now have students say the number sentence (for example, 5 and 2 make 7) as they show the 5-groups on their hands.

Note: Students see themselves improve as they continue working with numbers. Invite them to notice their improvement, and celebrate small successes and small steps toward mastery.

Note: The Fluency Practice activities have been shortened, assuming more time is needed for the culminating task.

Concept Development (40 minutes)

Preparation: Today's lesson incorporates elements of all of the Module 2 lessons. As in the culminating lesson for Module 1, students set up stations or exhibits displaying their knowledge of the shapes and solids they have studied. They represent each of the five shapes in a different display and associate geometric solids with the appropriate shapes, so there are at least five stations. To keep the number of students working at each station to a small group size, more stations can be added.

Materials: (T) Bag of flat shapes containing a triangle, a rectangle, a square, a circle, and a hexagon (as many as necessary in order to assign each group or pair of students one shape) (Lesson 5 Template) (S) Work mat (Activity Template); sets of cutouts from the module lessons; small pieces of modeling clay; geoboards and rubber bands; dot paper and markers; pattern blocks; paper and glue stick; crayons, pencils; Wikki Stix (if available); set of geometric solids; collection of pictures from catalogs, magazines, or newspapers representing the shapes and solids in real-life situations

T: We have been studying many flat shapes and solids. Today, you will get the chance to show what you know. We're having a Shape Fair!

T: I'm going to call you and your partner up to choose a shape from my bag. You will work with your partner to create an exhibit about that shape. You will go to a station and use the materials there to show what you know. You have cutouts, pictures, craft supplies, and drawing materials. What are some things you could do to show what you've learned about a shape?

S: We could draw it. → We could make it on a geoboard. → We could make it with our Wikki Stix! → We could have a shape hunt with the pictures. → We could cut it out.

T: There will be a bag of solids at your station as well. Could they help you to show your shape?

S: We can show the faces that match our shape. We could trace the solids.

T: Yes, some of the solids might show your shape, too! One last mission: I have a work mat for you to use. On the mat, the left side says, "These are (______)." The other side says, "These are not (______)." (Hold up the work mat, and demonstrate appropriate placement.) You will draw your shape in the blank spaces and then use this mat to do some sorting. Show things that do and don't match your shape in order to help your visitors understand the shape better.

MP.6

T: I will give you time to work on your exhibit, and after 20 minutes, I will give you a chance to visit the other exhibits in our Shape Fair. Students A, B, and C, please come choose your shape, and get started at your station.

Note: Use this time as an *informal assessment* tool for the close of the module. Circulate to observe student discussion and work. What representations are easiest and most familiar to the students? Are there some that might need review? What vocabulary and language do the students use in their discussions? Do they exhibit thorough understanding of the shapes and solids?

T: (When preparation time is up, allow students to rotate through the other exhibits.) Now, you may look at the rest of the Shape Fair. Talk with your partner about what you see at each station. What are the shapes and solids shown at each exhibit? How do you know? What ways did your friends choose to show them?

Suggestion: This would be a wonderful opportunity to have some other teachers, older students, or administrators come into the classroom to view the exhibits at the end of class. The students could explain their work to the visitors as an extension of the lesson.

Problem Set (0 minutes)

There is no Problem Set in this lesson to maximize available time for the culminating task.

Student Debrief (5 minutes)

Lesson Objective: Culminating task—collaborative groups create displays of different flat shapes with examples, non-examples, and a corresponding solid shape.

The Student Debrief is intended to invite reflection and active processing of the total lesson experience.

Any combination of the questions below may be used to lead the discussion.

- In what ways did you and your partner represent your shape?
- Which materials were easiest for you to use to explain your shape? Why?
- How did you decide which solids to use to represent your shape?
- Which shape(s) do you think were trickiest to make? Why?
- What new (or significant) math vocabulary did we use today to communicate precisely?

Name ______________________________ Date ______________

Shape Up Your Kitchen!

Search your kitchen to see what shapes and solids you can find. Make a kitchen-shaped collage by drawing the shapes that you see and by tracing the faces of the solids that you find. Color your collage.

Name ________________________________ Date _______________

These are (___).

These are not (___).

work mat

Student Name ___________________________

Topic A: Two-Dimensional Flat Shapes

Rubric Score: ___________ Time Elapsed: ___________

	Date 1	Date 2	Date 3
Topic A			
Topic B			
Topic C			

Materials: (S) Paper cutouts of typical triangles, squares, rectangles, hexagons, and circles; paper cutouts of variant shapes and difficult distractors (see Geometry Progression, p. 6)

1. (Hold up a rectangle. Use different shapes for each student.) Point to something in this room that is the same shape, and use your words to tell me all about it. How do you know they are the same shape?
2. (Place several typical, variant, and distracting shapes on the desk. Be sure to include three or four triangles.) Please put all the triangles in my hand. How can you tell they were all triangles?
3. (Hold up a rectangle.) How is a triangle different from this rectangle? How is it the same?
4. (Place five typical shapes in front of the student.) Put the circle next to the rectangle. Put the square below the hexagon. Put the triangle beside the square.

What did the student do?	What did the student say?
1.	
2.	
3.	
4.	

Topic B: Three-Dimensional Solid Shapes

Rubric Score: ___________ Time Elapsed: ____________

Materials: (S) 1 cone; 3 cylinders (wooden or plastic); a variety of real solid shapes (e.g., soup can, paper towel roll, party hat, ball, dice, or an unsharpened cylindrical—not hexagonal prism—pencil)

1. (Hand a cylinder to the student.) Point to something in this room that is the same solid shape, and use your words to tell me all about it.
2. (Place seven solid shapes in front of the student including three cylinders: wooden, plastic, and realia.) Put all the cylinders in this box.
3. (Show a cone.) How is the cylinder you are holding different from this cone? How is it the same?
4. (Place the set of solid shapes in front of the student.) Put the cube in front of the cylinder. Put the sphere behind the cone. Put the cone above the cube.

What did the student do?	What did the student say?
1.	
2.	
3.	
4.	

Topic C: Two-Dimensional and Three-Dimensional Shapes

Rubric Score: ___________ Time Elapsed: ___________

Materials: (T/S) Set of flat and solid shapes (do not use the paper cutouts from Topic A, but rather both commercial flat shapes and classroom flat shapes, such as a piece of colored construction paper, a CD sleeve, or a name tag)

1. Can you sort these shapes into one group of flat shapes and one group of solid shapes?
2. Tell me about your groups. What is the same about both groups? What is different?
3. Can you sort these shapes a different way? Tell me about your new groups. What is the same? What is different?

What did the student do?	What did the student say?
1.	
2.	
3.	

End-of-Module Assessment Task Standards Addressed	Topics A–C

Classify objects and count the number of objects in each category.

K.MD.3	Classify objects into given categories; count the numbers of objects in each category and sort the categories by count. (Limit category counts to be less than or equal to 10.)

Identify and describe shapes (squares, circles, triangles, rectangles, hexagons, cubes, cones, cylinders, and spheres).

K.G.1	Describe objects in the environment using names of shapes, and describe the relative positions of these objects using terms such as *above, below, beside, in front of, behind,* and *next to*.
K.G.2	Correctly name shapes regardless of their orientations or overall size.
K.G.3	Identify shapes as two-dimensional (lying in a plane, "flat") or three-dimensional ("solid").

Analyze, compare, create, and compose shapes.

K.G.4	Analyze and compare two- and three-dimensional shapes, in different sizes and orientations, using informal language to describe their similarities, differences, parts (e.g., number of sides and vertices/"corners") and other attributes (e.g., having sides of equal length).

Evaluating Student Learning Outcomes

A Progression Toward Mastery is provided to describe and quantify steps that illuminate the gradually increasing understandings that students develop *on their way to proficiency*. In this chart, this progress is presented from left (Step 1) to right (Step 4). The learning goal for students is to achieve Step 4 mastery. These steps are meant to help teachers and students identify and celebrate what the students CAN do now and what they need to work on next.

A Progression Toward Mastery				
Assessment Task Item	**STEP 1** **Little evidence of reasoning without a correct answer.** **(1 point)**	**STEP 2** **Evidence of some reasoning without a correct answer.** **(2 points)**	**STEP 3** **Evidence of some reasoning with a correct answer or evidence of solid reasoning with an incorrect answer.** **(3 points)**	**STEP 4** **Evidence of solid reasoning with a correct answer.** **(4 points)**
Topic A **K.G.1** **K.G.2** **K.G.4**	Student: ▪ Is unable to select, position, or describe indicated shapes. ▪ Takes considerable time to complete tasks, looks to the teacher for help often.	Student: ▪ Sorts indicated shapes randomly, resulting in some correct and some incorrect shapes in the group. ▪ Struggles to select, position, and describe indicated shapes.	Student: ▪ Identifies a shape from the environment but is unable to discuss its attributes. ▪ Sorts most of the indicated shapes. ▪ Correctly selects both of the indicated shapes but places them in the wrong position.	Student correctly: ▪ Identifies and describes several attributes of the shape from the environment that match the shape being shown to him. ▪ Sorts all indicated shapes from several typical, variant, and distracting shapes. ▪ Selects indicated shape and positions this shape below, next to, or beside another indicated shape.
Topic B **K.G.1** **K.G.2** **K.G.4**	Student: ▪ Is unable to select, position, or describe indicated shapes. ▪ Takes considerable time to complete tasks, looks to the teacher for help often.	Student: ▪ Sorts indicated solids randomly, resulting in some correct and some incorrect solids in the group. ▪ Struggles to select, position, and describe indicated solids.	Student: ▪ Identifies a solid from the environment but is unable to discuss its attributes. ▪ Sorts most of the indicated solids. ▪ Correctly selects both of the indicated solids but places them in the wrong position.	Student correctly: ▪ Identifies and describes several attributes of the solid from the environment that match the solid being shown to him. ▪ Sorts all indicated solids. ▪ Selects indicated solid and positions this solid above, in front of, or behind the indicated solid.

A Progression Toward Mastery				
Topic C **K.G.3** **K.MD.3**	Student: ▪ Incorrectly groups the shapes. ▪ Is not able to verbalize reasoning, or reasoning is not sound.	Student: ▪ Can sort the shapes into a group but is not able to verbalize reasoning. ▪ Cannot make a second grouping.	Student: ▪ Is able to sort the shapes into two groups but may or may not be able to verbalize reasoning. ▪ Is able to sort the shapes a second time but is unable to verbalize reasoning.	Student: ▪ Correctly sorts the shapes into two groups and is able to clearly state the reason the shapes belong to each group. ▪ Is able to sort the shapes again according to a different attribute and is able to state such an attribute.

Class Record Sheet of Rubric Scores: Module 2				
Student Names:	**Topic A:** Two-Dimensional Flat Shapes	**Topic B:** Three-Dimensional Solid Shapes	**Topic C:** Two-Dimensional and Three-Dimensional Shapes	**Next Steps:**

Answer Key

Eureka Math®
Grade K
Module 2

Special thanks go to the Gordon A. Cain Center and to the Department of Mathematics at Louisiana State University for their support in the development of *Eureka Math*.

Answer Key

GRADE K • MODULE 2

Two-Dimensional and Three-Dimensional Shapes

Lesson 1

Problem Set

4 shapes with curves placed on left side of chart; 10 shapes without curves placed on right side of chart

Fluency Template

1	1
2	2
3	3
4	2
1	2
2	1
3	1

Homework

Line drawn from square to checkerboard

Line drawn from rectangle to flag

Line drawn from circle to clock

Line drawn from triangle to sign

Line drawn from hexagon to parachute

Lesson 2

Problem Set

6 triangles colored blue; X placed on 9 shapes

Triangles drawn

Homework

4 triangles colored red; 5 shapes colored blue

2 triangles drawn

GK-M2-TE-B2-1.3.1-01.2016

Lesson 3

Problem Set

6 rectangles colored red; X placed on 8 shapes

Rectangles drawn

Homework

7 rectangles colored red; 7 triangles colored green

2 rectangles and 3 triangles drawn; 5

GK-M2-TE-B2-1.3.1-01.2016

Lesson 4

Problem Set

3 circles colored green; 5 hexagons colored yellow; X placed on 12 shapes

Hexagons and circles drawn

Homework

3 triangles colored blue; 4 rectangles colored red; 2 circles colored green; 4 hexagons colored yellow

2 triangles and 1 hexagon drawn; 3

Lesson 5

Problem Set

Square colored blue glued above duck

Circle colored yellow glued behind duck

Triangle colored green glued below duck

Rectangle colored red glued behind duck

Hexagon colored orange glued in front of duck

Circle with cutout colored purple and glued above duck

Homework

Blue square drawn behind elephant

Yellow circle drawn above the elephant

Green triangle drawn in front of elephant

Red rectangle drawn below elephant

Orange hexagon drawn below elephant

1 hexagon and 4 triangles drawn; 5

Lesson 6

Problem Set

Line drawn from party hat to cone

Line drawn from soccer ball to sphere

Line drawn from dice to cube

Line drawn from can to cylinder

Answers may vary.

Homework

Shapes pasted from magazine or drawn to match cylinder, sphere, cone, and cube

Lesson 7

Problem Set

3 cylinders circled red

2 cubes circled yellow (answer could also be 3 cubes circled yellow if the student circles each die individually)

3 cones circled green

3 spheres circled blue

Homework

1 cube pasted on left side of chart; 3 other shapes pasted on right side of chart

Answers will vary.

Lesson 8

Problem Set

Sphere pasted above train

Cube pasted behind train

Cylinder pasted in front of train

Cone pasted below train

Homework

Third car colored green

Second car circled blue

First car colored red

Road drawn below cars

Policeman drawn in front of cars

Sun drawn above cars

Lesson 9

Problem Set

6 flat shapes circled red

9 solid shapes circled green

Homework

Circle circled

Flat shape (hexagon) circled

Shape with straight side circled

Solid shape circled

Flat shape circled

Lesson 10

Homework

Answers may vary.